FOOD SCIENCE AND TECHNOLOGY

POLYPHENOLICS

FOOD SOURCES, BIOCHEMISTRY AND HEALTH BENEFITS

FOOD SCIENCE AND TECHNOLOGY

Additional books in this series can be found on Nova's website under the Series tab.

Additional e-books in this series can be found on Nova's website under the e-books tab.

FOOD SCIENCE AND TECHNOLOGY

POLYPHENOLICS

FOOD SOURCES, BIOCHEMISTRY AND HEALTH BENEFITS

PATRICIA CLARK
EDITOR

New York

NOTICE TO THE READER

Library of Congress Cataloging-in-Publication Data

ISBN: 978-1-53610-709-8

Published by Nova Science Publishers, Inc. † New York

CONTENTS

PREFACE

Polyphenolic compounds are part of a division of secondary metabolites with high antioxidant capacity, found naturally in most edible and inedible plants. These compounds are responsible for pigmentation and plants defense of pathogens agents. Its molecular structure includes phenolic hydroxyl groups and aromatic rings. Polyphenols are increasingly studied because of the benefits to human health, introducing the ability to eliminate free radicals generated by oxidation, reducing risk of cancer, anti-aging, anti-diabetic, among others. This book provides new research on food sources of polyphenolic compounds, and their biochemistry and health benefits.

Chapter 1 – The polyphenolic compounds are products of high importance due to the different applications in the medical, chemical, food and cosmetic industry. These compounds have many properties as antioxidant, anti-cancer, anti-diabetic and other biological effects. Many extractives are obtained from fruits and vegetables, being of great interest in the industry due to their antioxidant capacity, and nutritional value to food. However, there are no reports in literature related to the effect of the polyphenolic compounds purity in the economic profitability and their final application. In this chapter, the classification of the polyphenolic compounds and the required purities for their final application are presented. Different technologies for the extraction of polyphenolic compounds are discussed, including conventional and unconventional extraction. Two study cases of extraction are analyzed through simulation and economic analysis using Aspen Plus (Aspen Technology Inc.): chlorogenic acid (CGA) from spent coffee grounds and anthocyanins from blackberry pulp. It was possible to demonstrate that the purity (for different applications) has a strong influence on the profitability for these type of processes.

Chapter 2 – Polyphenols are secondary metabolites of plants, containing in their structure the aromatic ring with one or more phenolic groups. Such molecules have great antioxidant potential. They can modulate the activity of many enzymes and cell receptors and generate specific biological effects. Several polyphenols are already in use as active ingredients of pharmaceuticals or different cosmetic products. The large size of these polycyclic molecules and poor solubility in aqueous and/or lipid media may limit development of stable formulations and their passive diffusion into and through the biological membranes. The current strategy for enhancement of solubility and permeability of biomedically relevant polyphenols is based on formation of specific complexes with phospholipids which self associate in aqueous media and form specific unilamellar vesicles often described as phytosomes or herbosomes. There is an increasing number of scientific publications in this field during the last decade and different polyphenolics (e.g., curcuminoids, silybin, silymarin, ginkgoflavonglucosides, ginkgolides, bilobalide, and dimeric flavonoids from *Ginkgo biloba* leaf, sericoside, polyphenols from *Camellia sinensis* leaf (epigallocatechin, catechin, epicatechin-3-O-gallate, epigallo catechin-3-O-gallate), *Vitis Vinifera* (resveratrol, quercitin, catechin, procyanidins, epicatechin), are marketed as phytosomes obtained by the patented technology Phytosome® (Indena, Italy). The chapter reviews the significance of encapsulation of poliphenolic active ingredients, properties of commercially available phytosomes (physicochemical characteristics, stability, loading capacity, solubility, dispersibility) and phytosome relevance for development of pharmaceutical preparations and personal care products. The potential for enhancement of biological effects (cardiovascular, antiinflammatory, hepatoprotective, anticancer, hypoglycemic, weight loss, antiaging), of the currently investigated phytosomes is reviewed comprehensively.

Chapter 3 – *Moringa oleifera* Lam, a very resilient plant, currently cultivated in all the tropical and sub-tropical regions of the world has a rich history as traditional medicine and food. *M. oleifera* is being used to improve the nutritional value of staple foods in many parts of the world including Africa. It is considered as an important food forticant in view of the appreciable amounts of proteins, carbohydrates, starch as well as specific micronutrients like calcium, iron, magnesium, manganese, copper and potassium particularly in the leaves. Likewise, an impressive range of intrinsic bioactive phytonutrients with known antioxidant promises such as ascorbic acid, phenolics, flavonoids, tocopherols and carotenoids have also been recognized in this miracle tree. In addition, to its high nutritive values and powerful water purifying abilities, it is being considered for the prevention or treatment of a number of non-infectious ailments because of its pluripharmacological properties ranging from antioxidants, anti-inflammatory, anti-hypertensive to anticancer. These properties can be ascribed to the host of polyphenolic compounds present in these plants. Reported studies conducted on experimental animals and on human cell lines, though inadequate in number, seem concordant in their support for these properties. Therefore, the nutritional, prophylactic and therapeutic potentials of this multipurpose plant are being commended and this has allowed to envisage their potential applications as fresh produce or dried as a functional food and nutraceutical health promoter. This chapter will therefore, explore current scientific data on the therapeutic potential of the phenolic phytonutrients from *M. oleifera* plants in relation to the management of non-communicable diseases.

Chapter 4 – *Punica granatum* L. (pomegranate) has been valued for its medicinal properties since ancient times and is still being used in folk medicine across the world. The different anatomical parts of pomegranate have been professed with different health benefits in different indigenous cultures and associated traditional medicine. The dietary nature of pomegranate has attracted significant scientific interest at validating its ethnomedicinal uses as well as promoting its use as a functional food to mitigate chronic human diseases. The prophylactic effects of pomegranate have been attributed to its polyphenolic richness, which has shown potent antioxidant and anti-inflammatory capacities, both *in vitro* and *in vivo*. Oxidative stress and high inflammation have been reported as the underlying pathophysiological hallmarks of a number of chronic diseases and attenuating the inflammatory state and increasing the antioxidant status constitute an important target in managing the latter. Dietary pomegranate polyphenols have been shown to reduce plasma oxidative stress markers namely protein carbonyls and malondialdehyde as well as to modulate C-reactive protein, sE-selectin, TNF-α and IL-6. The *in vivo* pharmacological and therapeutic effect of pomegranate polyphenol consumption is mostly attributed to the bio-transformed metabolites of pomegranate bioactive polyphenols, following ingestion in the human body, as evidenced by increasing amount of literature available from pomegranate pharmacokinetic studies. This chapter, therefore, reviews findings on the bio-activity of pomegranate polyphenols and their bio-transformed metabolites with emphasis on their *in vivo* anti-oxidative and anti-inflammatory mechanisms of action. The chapter will conclude on the relevance and potentials of pomegranate polyphenols in clinical trials.

Chapter 5 – Worldwide mortality from infectious ailments is being substituted by chronic degenerative diseases, such as cancer, obesity, diabetes mellitus and neurological conditions. It is known that natural compounds may affect the expression and gene transcription which influence mechanisms involved in chronic diseases. Furthermore, they may also influence the predisposition to certain disorders as a result of individual genetic variability. For these reasons, nutrition research focuses not only on nutrient deficiency but also in the prevention of various progressive diseases. The inclusion of natural products such as fruit, beverages, teas and vegetables in the diet has long been related with various health benefits. These beneficial effects have been associated to the presence of polyphenolic compounds and their pharmacological properties. With regards to beneficial properties, some reports have revealed that polyphenolics could interact with cellular signaling pathways regulating the activity of transcription factors and affecting the expression of genes. Genes differentially expressed include genes involved in a wide range of physiological and pathological functions, such as metabolism, transport, enzyme activity, signal transduction or transcription. In this chapter, the authors review studies assessing modulation of genes expression by dietary polyphenolics that could constitute a new pathway by which these phytocompounds may exert their health effects.

In: Polyphenolics
Editor: Patricia Clark

ISBN: 978-1-53610-709-8
© 2017 Nova Science Publishers, Inc.

Chapter 1

ECONOMIC ASSESSMENT OF POLYPHENOLIC COMPOUNDS PRODUCTION AT DIFFERENT PURITIES AND APPLICATIONS

Ashley Sthefanía Caballero Galván,
Carlos Eduardo Orrego Alzate
and Carlos Ariel Cardona Álzate[*]
Instituto de Biotecnología y Agroindustria,
Departamento de Ingeniería Química,
Universidad Nacional de Colombia,
Manizales campus, Manizales, Colombia

ABSTRACT

The polyphenolic compounds are products of high importance due to the different applications in the medical, chemical, food and cosmetic industry. These compounds have many properties as antioxidant, anti-cancer, anti-diabetic and other biological effects. Many extractives are obtained from fruits and vegetables, being of great interest in the industry due to their antioxidant capacity, and nutritional value to food. However, there are no reports in literature related to the effect of the polyphenolic compounds purity in the economic profitability and their final application. In this chapter, the classification of the polyphenolic

---

[*] Corresponding Author: ccardonaal@unal.edu.co

compounds and the required purities for their final application are presented. Different technologies for the extraction of polyphenolic compounds are discussed, including conventional and unconventional extraction. Two study cases of extraction are analyzed through simulation and economic analysis using Aspen Plus (Aspen Technology Inc.): chlorogenic acid (CGA) from spent coffee grounds and anthocyanins from blackberry pulp. It was possible to demonstrate that the purity (for different applications) has a strong influence on the profitability for these type of processes.

1. INTRODUCTION

The polyphenolic compounds are part of a division of secondary metabolites with high antioxidant capacity and it is found naturally in most edible and inedible plants [1]. These compounds are responsible for pigmentation and plants defense of pathogens agents [2,3]. Its molecular structure includes phenolic hydroxyl groups and aromatic rings [4]. Polyphenols have been studied due the benefits to human health that they provide. These include the ability to eliminate free radicals generated by oxidation [5], reducing risk of cancer, anti-aging, anti-diabetic, among others. The above is shown in Figure 1 [6]. The polyphenolic compounds include a large group of chemical compounds that can be classified according to the number of carbon atoms present in four representative subgroups: flavonoids, phenolic acids, tannins and stilbenes. These are produced in different plant places [7].

Different studies have reported the polyphenolic compounds extraction from a great variety of raw materials and different technologies (see Table 1). The polyphenolic compounds obtainment requires the use of extraction methods. Extraction technologies are mainly characterized by the application of high temperatures, the requirement of the raw material particle size reduction, the longtime operation and low yield. However, the use of high temperatures can cause the compounds degradation due its temperature sensitivity. Alternative technologies have emerged in response to improving the polyphenolic compounds obtainment. Between these technologies can be mentioned the microwave assisted extraction, vacuum, ultrasound assisted extraction (UAE) and supercritical fluid extraction (SFE) [8].

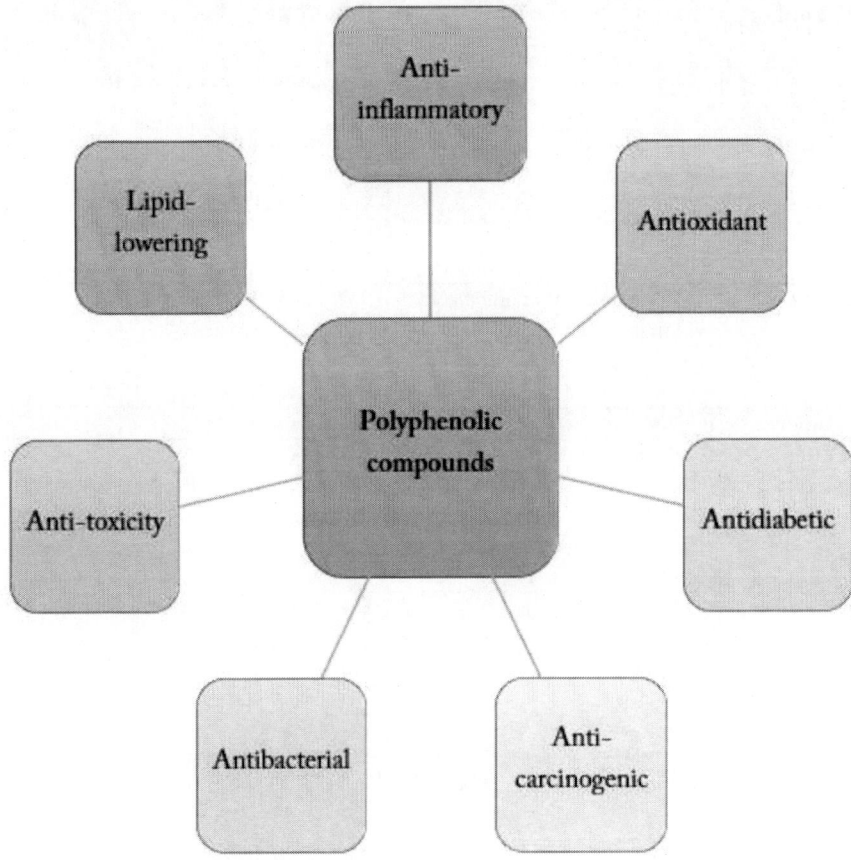

Figure 1. Characteristic of polyphenolic compounds.

Polyphenolic compounds have various applications related to the compounds obtained and the purity thereof. Products with high purities (i.e. with higher polyphenolic compounds content) are used as food preservatives and they are employed by the pharmaceutical and cosmetics industries. By other hand, products with low purity are used for the paper production, paints, natural colorings, essences [18]. For the concentration and purifying of polyphenolic compounds extracts are employed different technologies, among which the most representative are vacuum distillation and membrane technology, the first being the most conventional [19, 4].

Table 1. Raw materials used in the extraction of polyphenolic compounds

Raw Material	PolyphenolicCompound	ExtractionTechnology	References
Blackberry	Anthocyanins	Ultrasound assisted extraction	[9]
		Supercritical Fluid Extraction	[10]
Pear	Caffeicacid, Chlorogenicacid	Solventextraction	[11]
Peach	Caffeic acid, Chlorogenic acid	Solvent extraction	[11]
Apple	Phlorizin, Chlorogenic acid, Catechins, Cyanidin, Quercetin	Ultrasound assisted extraction	[12]
		Solvent extraction	[13]
Sunflower	Chlorogenic acid	Solvent extraction	[14]
Coffee and wastes	Gallic acid, Catechin, Cyanidin, Quercetin, Kaempferol, Epicatechin Vanillic acid, Ferulic acid, Chlorogenic acid	Solvent extraction Ultrasound assisted extraction	[15] [16]
Tree tomato	Cyanidin, Pelargonidin, Delphinidin, Chlorogenic acid	Solvent extraction	[17]
Naranjilla	Chlorogenic acid	Solvent extraction	[17]

2. PURITY AND APPLICATIONS

Polyphenolic compounds with different purities have a great variety of applications. Those required for pharmaceutical products and chemical compounds needs a purity of 90% or higher. Nevertheless, these have high costs when they are compared with the employed in food and essences that only requires purities close to 40-80%. Most polyphenolic compounds with high purity are obtained by extraction-purification, either by the use of vacuum distillation or membrane technology. Table 2 shows the polyphenolic compounds prices related to their respective purities and application.

3. EXTRACTION TECHNOLOGIES

Conventional and unconventional extraction techniques are used for the polyphenolic compounds extraction. Some conventional solvent extraction methods are soxhlet extraction, mechanical extraction [20]. Following the nonconventional definition the supercritical fluid extraction, ultrasonic assisted extraction and microwave-assisted extraction are some of this techniques [21, 22]. Below is a brief description of some conventional and unconventional technologies.

Table 2. Price of some polyphenolic compounds

Compound	Purity	Applications	Price (USD/g)
Gallic acid	Standard	Reference standard	962.00
	97.5-100%	Titration, HPLC	0.875
Chlorogenic acid	Standard	Pharmaceutical reference standard	7700.00
	≥95%	Titration	103.00
	55-45%	Food extract	37.125
	≤45%	Food capsules	17.31
Caffeic acid	≥99%	Chemical reactive, pharmaceutical, HPLC	95.80
	≥98%	HPLC	24.6
	≥95%	HPLC	8.5
B-carotene	≥97%	Chemical reactive, pharmaceutical	20.30
	≥93%	Chemical reactive, pharmaceutical	11.54
Vanillin acid	≥97	Chemical reactive	2.27
p-coumaric acid	≥98	Chemical reactive, pharmaceutical	12.30
Catechin	≥98%	Chemical reactive, pharmaceutical, HPLC	13.5
Salicylic acid	≥99%	Chemical reactive	0.199
	≤30%	Cosmetic	0.064
Ferulic acid	Standard	Pharmaceutical reference standard	1185
	≥99%	Chemical reactive, pharmaceutical	2.96
Flavone	≥98%	Chemical reactive	28.10
Rutin	Standard	Pharmaceutical reference standard	2372
	≥94%	HPLC	2.37
Sinapic acid	≥99	Chemical reactive, pharmaceutical	65.90
	≥98	Chemical reactive, pharmaceutical	58.40

3.1. Conventional Technologies

3.1.1. Solvent Extraction

Solvent extraction technology (SE) is the oldest technology for obtaining polyphenolic compounds. The process begins with the reception of raw materials that is usually dried. The stream with the raw material is sent to particle size reduction in mills to increase the raw material exposure into extraction the extraction process. Then the resulting stream enters an extraction column in which the solvent is added previously and then heated to the required operating system conditions (temperature and solvent-solid ratio). In the column it is obtained different output streams, one with solids, liquid solvent and extractives. The aqueous phase is carried out to a remaining solid matrix and liquids separation by filtration. The organic phase is sent to evaporation in order to recover the solvent and recycle it to the process. In Figure 3, the solvent extraction process flow diagram is observed. However, this technology has many disadvantages ranging from the requirement of long periods and high temperatures which can cause oxidation and further damage to the environment [23].

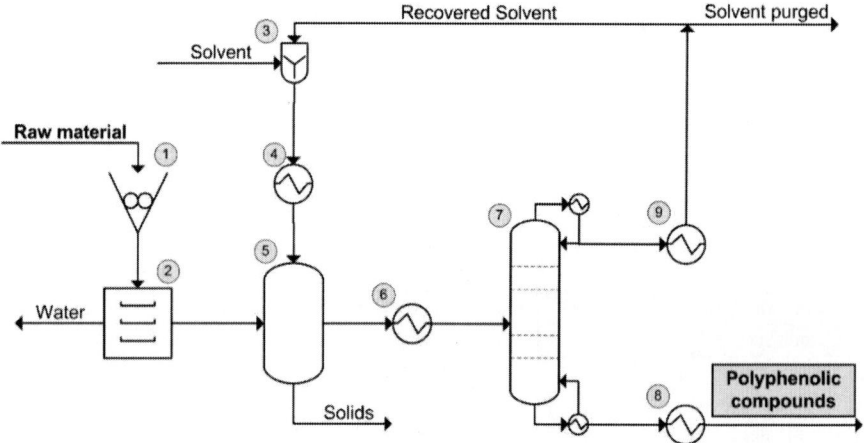

Figure 3. Flow diagram solvent extraction process.

1. Grynder, 2. Dryer, 3. Mixer, 4-6-8-9. Heat exchanger, 5. Extraction column, 7. Evaporation Column.

3.1.2. Soxhlet Extraction

This technology perform extractions automatically using a solvent which evaporates and condenses [24]. The solid sample is placed in a thimble which is led to the soxhlet extraction chamber. The process is carried at boiling temperature continuously to exhaust the sample and obtaining the extractives in the volume ball [25]. This technology has many disadvantages given the extended processing time and high temperatures.

3.2. Nonconventional Technologies

The non-conventional technologies implementation can improve the extraction of polyphenolic compounds. Technologies such as supercritical fluid extraction (SFE), ultrasound assisted extraction (UAE) and microwave assisted extraction (MAE), have great advantages over conventional technologies due the environmental impact reduction by using less aggressive solvents, lower extration times and higher performance.

3.2.1. Ultrasound Assisted Extraction

Ultrasound assisted extraction is a process that has great advantages over conventional techniques due they are a friendly technology that is more efficient and have low cost of implementation and requirements for equipment due to the simplicity of method development [21]. Ultrasound is applied in solid or liquid media using as fluids a gas or liquid. This has a variety of applications among which are the cleaning, sterilization, chemical reactions acceleration, pretreatment, oxidation, extraction, and so forth. [26, 27]. Different processes using ultrasound include a frequency generator 18 KHz to 100 MHz, a transducer and a reactor.

The process involves a sound transmission (sonication) element through a medium generating the vibration of molecules through waves that create cavitation in the solution. Increasing pressure and temperature generated by cavitation, destroys the walls of the matrix raw material, releasing the polyphenolic compounds in the solution [28].

3.2.2. Supercritical Fluids Extraction

The supercritical fluid extraction (SFE) is a technology of mass transfer upper than the solvent critical point as shown in Figure 3. This extraction technology has great advantages over conventional techniques given its high selectivity and efficiency. The SFE has a wide range of applications in pharmaceutical, food, oil industries and waste treatments [29]. The SFE uses the solubilization ability of a fluid having properties of temperature and pressure above the critical values to perform a more optimal work and produce changes in the density and solvent power in terms of separation efficiency [30]. This technology uses low temperatures in absence of air, preventing oxidation and thermolabile components degradation [9].

Different works have employed solvents such as water, ethanol, methanol, carbon dioxide, acetone, among others for supercritical fluid extraction. The most used solvent for this type of extraction is carbon dioxide (CO_2) due to its low critical temperature and critical pressure (3°C and 73.8 bar). Additionally carbon dioxide is non-toxic, easily recycled, nonflammable, chemically inert and has low cost [31]. The CO_2 has characteristics for safe removal of the interest components in the plant material. It is not present in the spent solids due to solvent evaporation at environmental temperature and pressure.

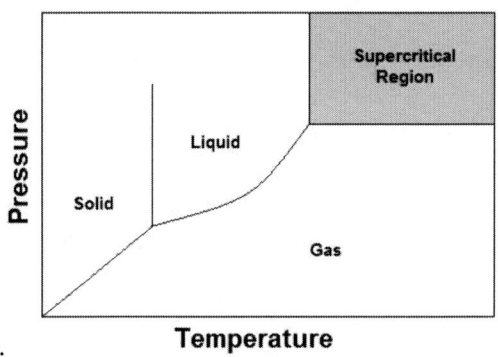

Figure 3. Region of supercritical fluids.

3.2.3. Microwave Assisted Extraction

Microwave assisted extraction (MAE) is a heating method with radiation causing the chemical bonds break of the molecules [28]. In the Figure 4, the process diagram for polyphenolic compounds extraction is presented. The MAE breaks the plant tissue sample, releasing the polyphenolic compounds in the solvent [32].

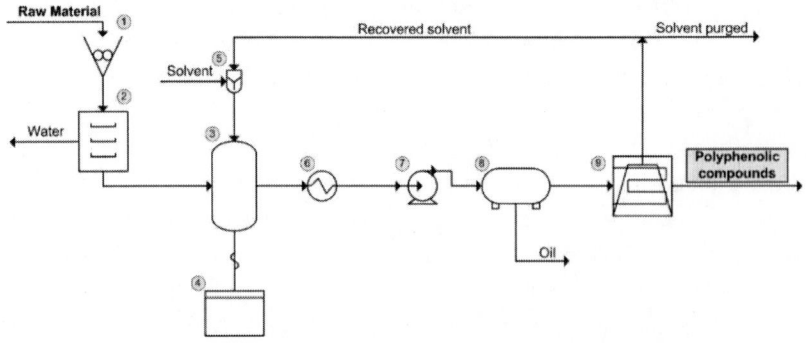

Figure 4. Flow diagram of extraction of polyphenolic compounds with MAE.

1. Grinder, 2. Dryer, 3. Extractor, 4. Microwave, 5. Mixer, 6. Heat exchanger, 7. Pump, 8. Decanter, 9. Evaporator.

4. METHODOLOGY

4.1. Process Description

Using computational tools, the simulation processes is performed for different cases. The first case is the Chlorogenic Acid extraction from Spent Coffee Grounds and the second case is the Anthocyanin extraction from Blackberry Pulp. Aspen Plus (Aspen Technology Inc., USA) simulation tool was used to specify the raw materials transformation taking into account, as main criterion, to obtain different purities of the interest product (i.e. 50, 75 y 95%). In addition, design parameters, chemical and physical properties were used to calculate the mass and energy balances for each process. As an extraction technology SFE is used in two cases to be described below.

4.1.2. Case 1: Chlorogenic Acid from Spent Coffee Grounds

Chlorogenic acid (CGA) is a thermosensitive polyphenol with high added value due to its uses as an antioxidant, hypoglycemic, antidiabetic anti-inflammatory, anti-aging and other biological effects [33, 34, 35]. The chlorogenic acid is present in fruits, vegetables and coffee. CGA extraction from coffee grounds gives an extract with a composition shown in Table 3.

Table 3. Composition of spent coffee grounds [36, 37]

Compound	Percentage (%)	Compound	Percentage (%)
Moisture	0.990	Gallic acid	1.249
Cellulose	8.911	Chlorogenic acid	2.798
Arabine	1.188	Catechin	0.300
Galacton	13.663	Caffeic acid	0.035
Mannan	13.168	Rutin	0.100
Protein	2.772	Ellagic acid	0.100
Lignin	29.505	P-cumaric acid	0.015
Ash	2.277	Ferulic acid	0.003
Acetyl	2.574	Quercetin	0.350

Figure 5 shows the chlorogenic acid extraction scheme using CO_2 in supercritical conditions. As an initial step, the solid material is dried at a temperature of 40°C. The supercritical fluid used (carbon dioxide) previously is pressurized to get a liquid and then to reach supercritical conditions. The pressurized liquid CO_2 is carried to the extraction chamber in which previously was deposited the coffee grounds. Moreover, it is added the Pressurized cosolvent (ethanol). The cosolvent is added to the sample in order to increase the carbon dioxide polarity. Once the extraction is carried out to the system and it is depressurized by the use of a valve and a collector, which operates at a pressure less than the chamber. The collector separates the solids and CO_2 is recirculated. The extract obtained is separated into ethanol which is used by the system recirculation means. Finally, chlorogenic acid is obtained at the purity required for subsequent steps.

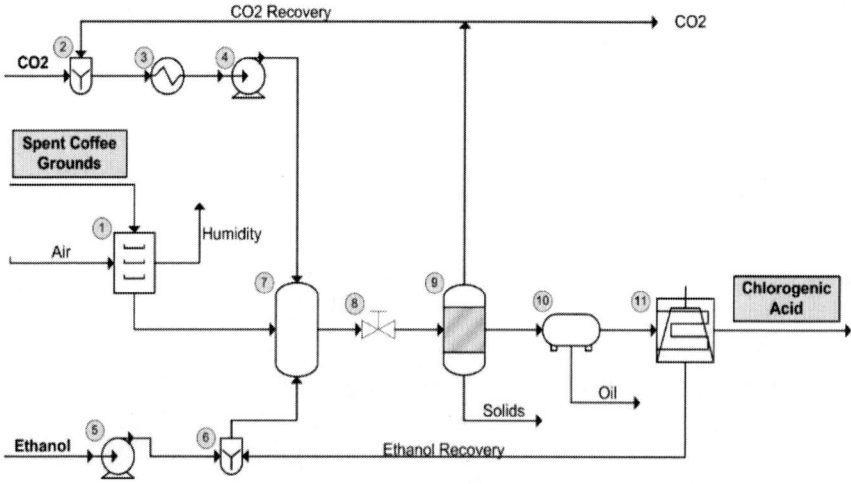

Figure 5. Flow Diagram of obtaining chlorogenic acid.

1. Dryer, 2. Mixer, 3. Heat exchanger, 4. Carbon dioxide pump, 5. Ethanol Pump, 6. Mixer, 7. Extractor, 8. Valve, 9. Collector, 10. Decanter, 11. Separation.

4.1.3. Case 2: Anthocyanin from Blackberry Pulp

The technology consists of operating at T and P above of the critical point of the solvent used (40°C, 300 bar), obtaining an extract rich stream with a solvent, and separating the solids from the stream. The system is taken to a depressurization and subsequent solvent recovery step, where is performed a pressure drop, decreasing the solubility and separating the product (extract). The recovered solvent is recycled to the initial stage realizing an increased pressure and cooling, as shown in Figure 6 [38]. In Table 4, the composition of the blackberry pulp used for this analysis is presented.

Table 4. Composition of blackberry pulp [10]

Compound	Percentage (%)
Moisture	5.12
Cellulose	41.74
Hemicellulose	19.8
Lignin	19.22
Ash	1,14
Anthocyanins	12.56
Other extractives	0.42

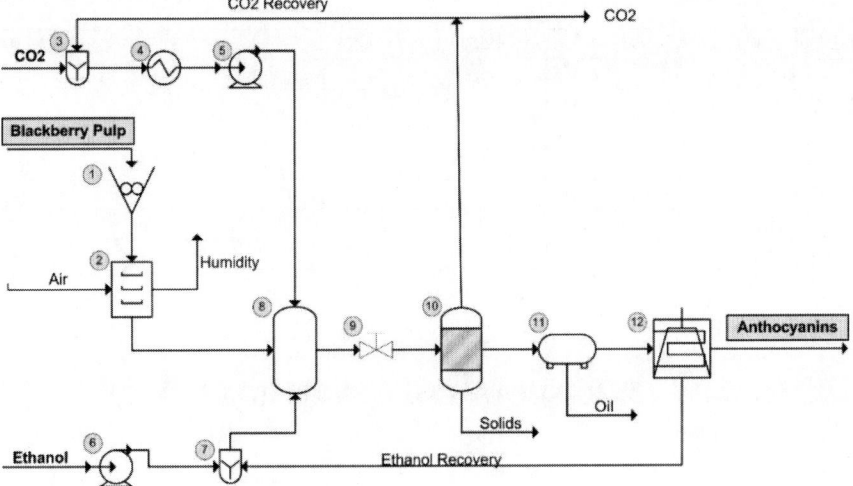

Figure 6. Flow Diagram of obtaining anthocyanins.

1. Grinder, 2. Dryer, 3. Mixer, 4. Heat exchanger, 5. Carbon dioxide pump, 6. Ethanol Pump, 7. Mixer, 8. Extractor, 9. Valve, 10. Collector, 11. Decanter, 12. Separation.

4.2. Economic Assessment

The development of the economic analysis was performed with the use of Aspen Process Economic Analyzer V8.2 (ASPEN TECHNOLOGY INC) software. As input data are supplied the annual interest rate, interest rate, utility costs, operating costs and raw materials. The above information is presented in Table 5. Finally, the straight-line method of depreciation is used over a period of 12 years.

Table 5. Investment parameters and prices used in the economic analysis

Item	Unit	Value	Reference
Investment Parameters			
Tax rate	%	25	[39]
Annual Interest rate	%	17	
Raw materials			
Spent coffee grounds	USD/kg	0.25	
Blackberry pulp	USD/kg	0.49	[40]
Carbon dioxide	USD/kg	1.55	
Ethanol	USD/kg	1.24	[41]
Item	Unit	Value	Reference
Utilities			
LP steam	USD/tonne	1.57	
MP steam	USD/tonne	8.18	[42]
HP steam	USD/tonne	9.86	
Potable water	USD/m^3	1.25	
Fuel	USD/MMBTU	7.21	[39]
Electricity	USD/kWh	0.10	
Operation			
Operator	USD/h	2.14	[39]
Supervisor	USD/h	4.29	

5. RESULTS AND DISCUSSIONS

5.1. Case 1: Chlorogenic Acid from Spent Coffee Grounds

The yields in the extraction of chlorogenic acid to the spent coffee grounds fed were 0.032, 0.023 and 0.018 kg product/kg raw material at purities of 50%, 75% and 95%, respectively. Production costs through supercritical fluid extraction of chlorogenic acid from spent coffee grounds are directly related to the extract purity as shown in Table 6. In which chlorogenic acid obtaining a purity of 95% has an additional cost in raw materials of 51.7% and 28.7% compared with a product of 50% purity and 75%, respectively.

Figure 7 shows the costs distribution of obtaining different purities chlorogenic acid (50, 75 and 95%). Raw material costs have the greatest influence on process costs more than 42% for each of the products obtained. Additionally, it is noted that operating costs and charges have low participation in the total costs of the process with just 2%.

Table 6. Cost of obtaining chlorogenic acid

	Purity of 50% (USD/kg)	Purity of 75% (USD/kg)	Purity of 95% (USD/kg)
Raw Materials	2.71	4.00	5.61
Operating Labor Cost	0.26	0.38	0.54
Maintenance Cost	0.25	0.36	0.47
Utilities	0.12	0.28	0.51
Operating Charges	0.06	0.09	0.13
Plant Overhead	0.25	0.37	0.50
General and Administrative Costs	0.29	0.44	0.55
Depreciation Expense	2.51	3.55	4.83

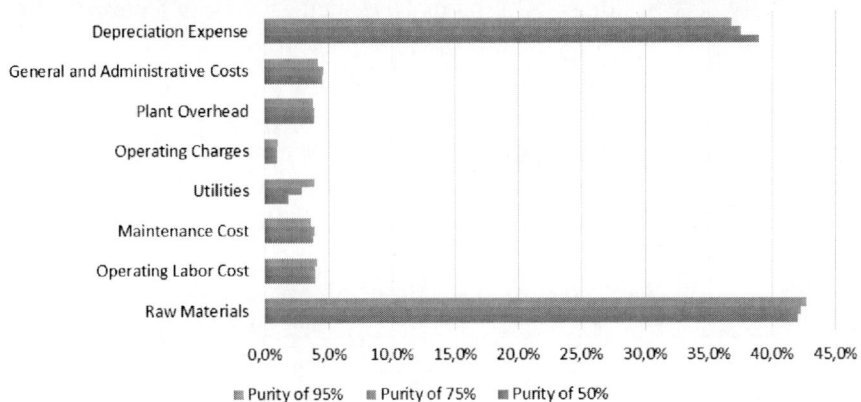

Figure 7. Distribution costs for chlorogenic acid extraction using SFE.

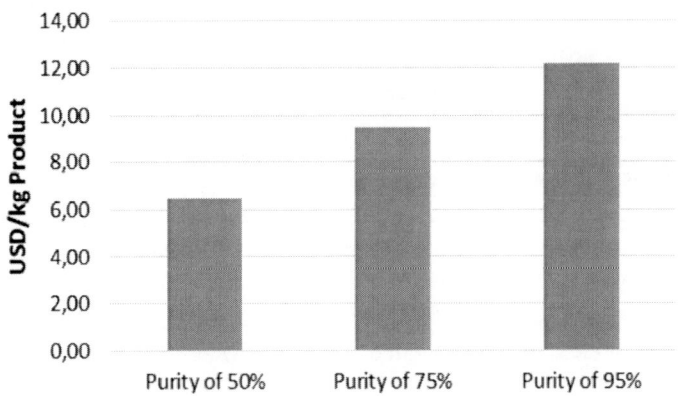

Figure 8. Total production cost per kg of extract of chlorogenic acid.

The total production cost was calculated taking into account the obtained extract flow in the process from 1.0 tonne/h of fed raw material: 32.57kg/h (purity of 50%), 22.89 kg/h (purity of 75%) and 17,753 kg/h (purity of 95%). Total cost of the process as a higher value for obtaining higher extract purities of 95% with a ratio of 0.5 and 0.27 with respect to purities of 50% and 75% was presented, as shown in Figure 8.

5.2. Case 2: Anthocyanin from Blackberry Pulp

Obtaining anthocyanins extract gave as a yields results 0.145, 0.108 y 0.098 kg product/kg raw material from 50%, 75% and 95% purity respectively. The yield for 50% purity, due to the impurities present in the product which results in a greater flow of the process.

The economic results of the extraction of anthocyanins from blackberry pulp showed that the purity factor has a great importance in process costs presenting increases up to twice its value as shown in Table 7. For products required for applications in the pharmaceutical and chemical industry (95% purity) compared to essences and food uses (50% purity).

Table 7. Cost of obtaining anthocyanins

	Purity of 50% (USD/kg)	Purity of 75% (USD/kg)	Purity of 95% (USD/kg)
Raw Materials	0.86	1.23	2.07
Operating Labor Cost	0.06	0.08	0.12
Maintenance Cost	0.04	0.05	0.08
Utilities	0.03	0.04	0.06
Operating Charges	0.01	0.02	0.03
Plant Overhead	0.05	0.07	0.10
General and Administrative Costs	0.09	0.12	0.17
Depreciation Expense	0.44	0.62	0.91

As shown in Figure 9, the production cost is mainly related to the raw materials costs (CO_2, ethanol and blackberry pulp) and depreciation expense as a share about 83%. Additionally, the same behavior is showed in the distribution of each of the evaluated costs for different purities.

As final cost of the processes, higher values for purity of 95% were obtained with an increase of 55% and 36% with respect to purities of 50% and 75% respectively. However, the price of products with higher purity has a high compensation for this business.

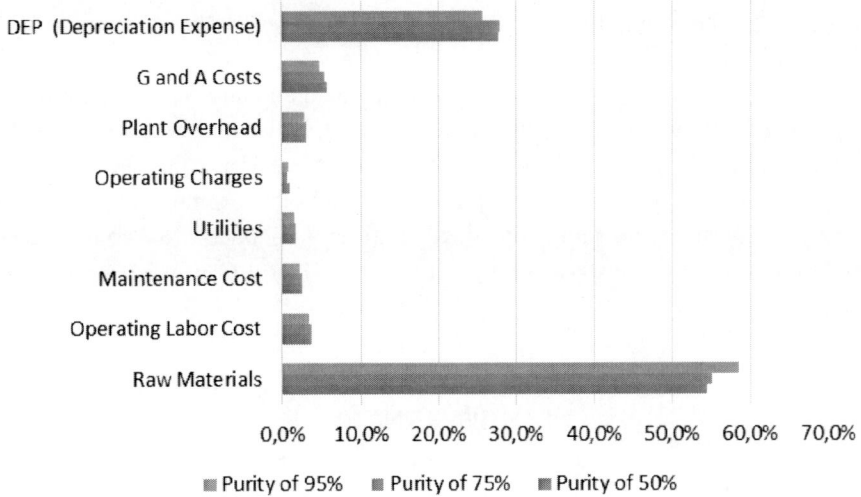

Figure 9. Distribución de costos de extracción de antocianinas.

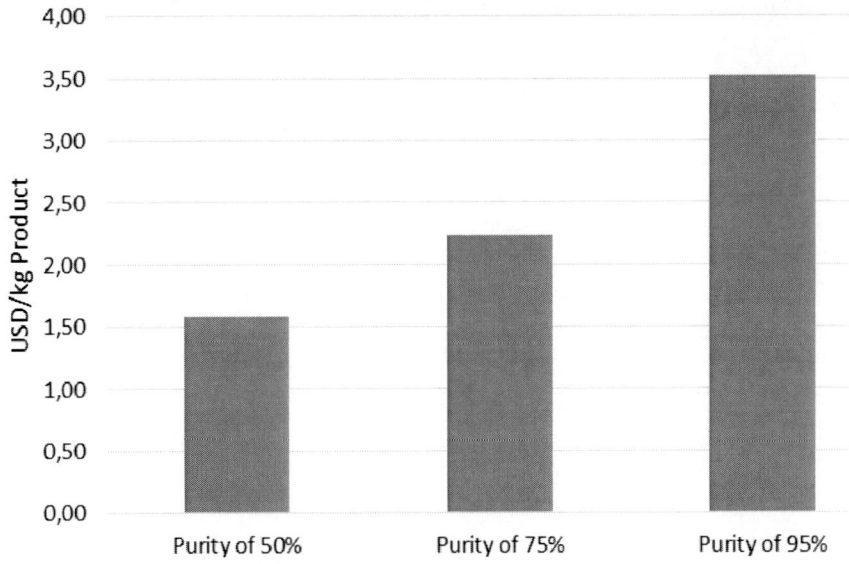

Figure 10. Total production cost per kg of extract anthocyanins.

CONCLUSION

The unconventional technologies use for obtaining polyphenolic compounds provides a great alternative to increase the benefits of the process, among which, higher performance, greater selectivity and better solubilization capacity can be found. In the case of SFE high purities can be achieved but the yields are reduced.

The total production costs of chlorogenic acid and anthocyanins presented a great increase when higher purities are reached. However, these costs give an interesting flexibility based on the very expensive extracts that are produced (high purity polyphenolic compounds can be twice or more expensive in comparison to conventionally low purities obtained in conventional technologies). So it is possible to get very good profits in a more compact equipment to get high added value products.

REFERENCES

[1]   F. Weber, N. Schulze-Kaysers, and A. Schieber, *Characterization and Quantification of Polyphenols in Fruits*. Elsevier, 2014.

[2]   T. Ozcan, a. Akpinar-Bayizit, L. Yilmaz-Ersan, and B. Delikanli, "Phenolics in Human Health," *Int. J. Chem. Eng. Appl.*, vol. 5, no. 5, pp. 393–396, 2014.

[3]   P. Maisuthisakul, "Use of Plant Phenolic Compounds as Antioxidants," pp. 222–238.

[4]   I. X. Cerón, R. T. L. Ng, M. El-Halwagi, and C. A. Cardona, "Process synthesis for antioxidant polyphenolic compounds production from Matisia cordata Bonpl. (zapote) pulp," *J. Food Eng.*, vol. 134, pp. 5–15, 2014.

[5]   S. Quideau, D. Deffieux, C. Douat-Casassus, and L. Pouységu, "Plant polyphenols: Chemical properties, biological activities, and synthesis," *Angew. Chemie - Int. Ed.*, vol. 50, no. 3, pp. 586–621, 2011.

[6]   G. A. and G. N and M. Akram, "Flavonoids and phenolic acids: Role and biochemical activity in plants and human," *J. Med. Plants Res.*, vol. 5, no. 31, pp. 6697–6703, 2011.

[7]   S. Martins, S. I. Mussatto, G. Martínez-Avila, J. Montañez-Saenz, C. N. Aguilar, and J. A. Teixeira, "Bioactive phenolic compounds: Production and extraction by solid-state fermentation. A review," *Biotechnol. Adv.*, vol. 29, no. 3, pp. 365–373, 2011.

[8]   W. HU, T. GUO, W.-J. JIANG, G.-L. DONG, D.-W. CHEN, S.-L. YANG, and H.-R. LI, "Effects of ultrahigh pressure extraction on yield and antioxidant activity of chlorogenic acid and cynaroside extracted from flower buds of Lonicera japonica," *Chin. J. Nat. Med.*, vol. 13, no. 6, pp. 445–453, 2015.

[9]   T. Bin Zou, M. Wang, R. Y. Gan, and W. H. Ling, "Optimization of ultrasound-assisted extraction of anthocyanins from mulberry, using response surface methodology," *Int. J. Mol. Sci.*, vol. 12, no. 5, pp. 3006–3017, 2011.

[10] J. Dávila, "Design of Biorefineries for High Value Added Products from Fruits Diseño de Biorefinerias para Productos de Alto Valor Agregado a partir de Frutas," Universidad Nacional de Colombia Sede Manizales, 2015.

[11] M. Carbonaro, M. Mattera, S. Nicoli, P. Bergamo, and M. Cappelloni, "Modulation of antioxidant compounds in organic vs conventional fruit (peach, Prunus persica L., and pear, Pyrus communis L.)," *J Agric Food Chem*, vol. 50, no. 19, pp. 5458–5462, 2002.

[12] M. A. Awad, A. De Jager, and L. M. Van Westing, "Flavonoid and chlorogenic acid levels in apple fruit: Characterisation of variation," *Sci. Hortic. (Amsterdam).*, vol. 83, no. 3–4, pp. 249–263, 2000.

[13] L. Jakobeka and A. R. Barronb, "Ancient apple varieties from Croatia as a source of bioactive polyphenolic compounds," *J. Food Compos. Anal.*, vol. 45, pp. 9–15, 2016.

[14] S. González-Pérez, K. B. Merck, J. M. Vereijken, G. A. Van Koningsveld, H. Gruppen, and A. G. J. Voragen, "Isolation and characterization of undenatured chlorogenic acid free sunflower (Helianthus annuus) proteins," *J. Agric. Food Chem.*, vol. 50, no. 6, pp. 1713–1719, 2002.

[15] Y. Sapozhnikova, "Development of liquid chromatography–tandem mass spectrometry method for analysis of polyphenolic compounds in liquid samples of grape juice, green tea and coffee," *Food Chem.*, vol. 150, pp. 83–93, 2014.

[16]   N. Abdullah Al-Dhabia, K. Ponmurugana, and P. Maran Jeganathanb, "Development and validation of ultrasound-assisted solid-liquid extraction of phenolic compounds from waste spent coffee grounds," *Ultrason. Sonochem.*, vol. 34, pp. 206–213, 2017.

[17]   C. Mertz, A. L. Gancel, Z. Gunata, P. Alter, C. Dhuique-Mayer, F. Vaillant, A. M. Perez, J. Ruales, and P. Brat, "Phenolic compounds, carotenoids and antioxidant activity of three tropical fruits," *J. Food Compos. Anal.*, vol. 22, no. 5, pp. 381–387, 2009.

[18]   R. Kushwaha and S. Karanjekar, "Standardization of ashwagandharishta formulation by TLC method," *Int. J. ChemTech Res.*, vol. 3, no. 3, pp. 1033–1036, 2011.

[19]   P. Chumsri, A. Sirichote, and A. Itharat, "Studies on the optimum conditions for the extraction and concentration of roselle (Hibiscus sabdariffa Linn.) extract," *Songklanakarin J. Sci. Technol.*, vol. 30, no. SUPPL. 1, pp. 133–139, 2008.

[20]   I. A. Saleh, M. Vinatoru, T. J. Mason, N. S. Abdel-Azim, E. A. Aboutabl, and F. M. Hammouda, "A possible general mechanism for ultrasound-assisted extraction (UAE) suggested from the results of UAE of chlorogenic acid from Cynara scolymus L. (artichoke) leaves," *Ultrason. Sonochem.*, vol. 31, pp. 330–336, 2016.

[21]   M. T. Mazvimba, Y. Yu, Z. Q. Cui, and Y. Zhang, "Optimization and orthogonal design of an ultrasonic-assisted aqueous extraction process for extracting chlorogenic acid from dry tobacco leaves," *Chin. J. Nat. Med.*, vol. 10, no. 4, pp. 311–320, 2012.

[22]   B. Zhang, R. Yang, and C. Z. Liu, "Microwave-assisted extraction of chlorogenic acid from flower buds of Lonicera japonica Thunb.," *Sep. Purif. Technol.*, vol. 62, no. 2, pp. 480–483, 2008.

[23]   X.-M. Li, S.-L. Tian, Z.-C. Pang, J.-Y. Shi, Z.-S. Feng, and Y.-M. Zhang, "Extraction of Cuminum cyminum essential oil by combination technology of organic solvent with low boiling point and steam distillation," *Food Chem.*, vol. 115, no. 3, pp. 1114–1119, 2009.

[24]   U. Morales, L. Alamilla, and R. Mora, "Extracción de compuestos de interés," *Agrowaste*, p. 7, 2013.

[25]   C. E. Núñez, "Extracciones Con Equipo Soxhlet," pp. 1–5, 2008.

[26]   C. Bendicho and I. LAvilla, "Ultrasound Extractions," *J. Chromatogr.*, pp. 1448–1454, 2000.

[27]   L. V. Daza Serna, "Assessment of Nonconventional Pretreatments for Agriculture Wastes Utilization," Universidad Nacional de Colombia, 2015.

[28] S. Q. Liew, G. C. Ngoh, R. Yusoff, and W. H. Teoh, "Sequential ultrasound-microwave assisted acid extraction (UMAE) of pectin from pomelo peels," *Int. J. Biol. Macromol.*, vol. 93, pp. 426–435, 2016.

[29] T. Knez, E. Markočič, M. Leitgeb, M. Primožič, M. K. Hrnčič, and M. Škerget, "Industrial applications of supercritical fluids: A review," *Energy*, vol. 77, pp. 235–243, 2013.

[30] H. Li, S. Li, and B. Shen, "Correlating and predicting the solubilities of solid n-alkanes in supercritical ethane using carnahan-starling-van der waals model," *Chinese J. Chem. Eng.*, vol. 21, no. 12, pp. 1360–1369, 2013.

[31] C. A. Cardona, L. Carlos, and P. Solano, *Introducción a las operaciones de separación no convencionales*, Primera. Manizales, 2007.

[32] M. Dhobi, V. Mandal, and S. Hemalatha, "Optimization of microwave assisted extraction of bioactive flavonolignan-silybinin.," *J. Chem. Metrol.*, vol. 3, no. 1, pp. 13–23, 2009.

[33] C. Miura, H. Li, H. Matsunaga, and J. Haginaka, "Molecularly imprinted polymer for chlorogenic acid by modified precipitation polymerization and its application to extraction of chlorogenic acid from Eucommia ulmodies leaves," *J. Pharm. Biomed. Anal.*, vol. 114, pp. 139–144, 2015.

[34] A. S. P. Moreira, M. A. Coimbra, F. M. Nunes, C. P. Passos, S. A. O. Santos, A. J. D. Silvestre, A. M. N. Silva, M. Rangel, and M. R. M. Domingues, "Chlorogenic acid–arabinose hybrid domains in coffee melanoidins: Evidences from a model system," *Food Chem.*, vol. 185, pp. 135–144, 2015.

[35] Z. Tan, C. Wang, Y. Yi, H. Wang, M. Li, W. Zhou, S. Tan, and F. Li, "Extraction and purification of chlorogenic acid from ramie (Boehmeria nivea L. Gaud) leaf using an ethanol/salt aqueous two-phase system," *Sep. Purif. Technol.*, vol. 132, pp. 396–400, 2014.

[36] E. E. Kwon, H. Yi, and Y. J. Jeon, "Sequential co-production of biodiesel and bioethanol with spent coffee grounds," *Bioresour. Technol.*, vol. 136, pp. 475–480, 2013.

[37] D. M. López-Barrera, K. Vázquez-Sánchez, M. G. F. Loarca-Piña, and R. Campos-Vega, "Spent coffee grounds, an innovative source of colonic fermentable compounds, inhibit inflammatory mediators in vitro," *Food Chem.*, vol. 212, pp. 282–290, 2016.

[38]   C. A. Cardona, J. A. Posada, and J. A. Quintero, "La Ingeniería Química Aplicada al Desarrollo de las Cadenas Agroindustriales y sus Residuos en Colombia," in *Aprovechamiento de subproducyos y residuos agroindustriales: Glicerina y Lignocelulósicos*, Primera., C. A. Cardona, J. A. Posada, and J. A. Quintero, Eds. Manizales, 2010, pp. 13–51.

[39]   J. A. Dávila, V. Hernández, E. Castro, and C. A. Cardona, "Economic and environmental assessment of syrup production. Colombian case," *Bioresour. Technol.*, vol. 161, pp. 84–90, 2014.

[40]   CORABASTOS, "Boletin Diario de Precios," 2016. [Online]. Available: http://www.corabastos.com.co/sitio/historicoApp2/reportes/BoletinDesc arga.php.

[41]   Fedebiocombustibles, "Etanol anhidro de caña. Cifras Informativasdel Sector Biocombustibles," 2011. [Online]. Available: http://www.fedebio combustibles.com.

[42]   J. Moncada, V. Hernández, Y. Chacón, R. Betancourt, and C. A. Cardona, "Citrus Based Biorefineries," in *Citrus Fruits. Production, Consumption and Health Benefits*, D. Simmons, Ed. Nova Publishers, 2015, pp. 1–26.

In: Polyphenolics   ISBN: 978-1-53610-709-8
Editor: Patricia Clark   © 2017 Nova Science Publishers, Inc.

Chapter 2

PROPERTIES AND BIOMEDICAL RELEVANCE OF PHYTOSOME ENCAPSULATED POLYPHENOLICS

Ljiljana Djekic and Danina Krajisnik*

University of Belgrade, Faculty of Pharmacy,
Department of Pharmaceutical Technology and Cosmetology,
Belgrade, Serbia

ABSTRACT

Polyphenols are secondary metabolites of plants, containing in their structure the aromatic ring with one or more phenolic groups. Such molecules have great antioxidant potential. They can modulate the activity of many enzymes and cell receptors and generate specific biological effects. Several polyphenols are already in use as active ingredients of pharmaceuticals or different cosmetic products. The large size of these polycyclic molecules and poor solubility in aqueous and/or lipid media may limit development of stable formulations and their passive diffusion into and through the biological membranes. The current strategy for enhancement of solubility and permeability of biomedically relevant polyphenols is based on formation of specific complexes with phospholipids which self associate in aqueous media and form specific unilamellar vesicles often described as phytosomes or herbosomes. There is an increasing number of scientific publications in this field during the

---

* E-mail: ljiljanadjek@gmail.com.

last decade and different polyphenolics (e.g., curcuminoids, silybin, silymarin, ginkgoflavonglucosides, ginkgolides, bilobalide, and dimeric flavonoids from *Ginkgo biloba* leaf, sericoside, polyphenols from *Camellia sinensis* leaf (epigallocatechin, catechin, epicatechin-3-O-gallate, epigallo catechin-3-O-gallate), *Vitis Vinifera* (resveratrol, quercitin, catechin, procyanidins, epicatechin), are marketed as phytosomes obtained by the patented technology Phytosome® (Indena, Italy). The chapter reviews the significance of encapsulation of poliphenolic active ingredients, properties of commercially available phytosomes (physicochemical characteristics, stability, loading capacity, solubility, dispersibility) and phytosome relevance for development of pharmaceutical preparations and personal care products. The potential for enhancement of biological effects (cardiovascular, antiinflammatory, hepatoprotective, anticancer, hypoglycemic, weight loss, antiaging), of the currently investigated phytosomes is reviewed comprehensively.

Keywords: polyphenolics, phospholipids, phytosome, pharmaceuticals, personal care products

INTRODUCTION

Plant derived substances with biomedical relevance have gained immense popularity and access to the global market as safer and effective alternative of synthetic ingredients of pharmaceutical and cosmetic products, which are considered to be full of adverse and potential toxic interactions. Currently, as many as one-third to approximately one-half of all the drugs available are derived from plants or other natural sources (Barnes et al., 2002; Khan et al., 2013; WHO, 2004). Polyphenols have received tremendous attention as pharmaceutical and cosmetic active ingredients. It is a highly diverse group of natural phenolic compounds comprising several sub-groups: phenolic acids, flavonoids (isoflavones, neoflavonoids, chalcones, flavones, flavonols, flavanones, flavanonols, flavanols, proanthocyanidins, anthocyanidins), polyphenolic amides, and non-flavonoid polyphenols (e.g., resveratrol, curcumin) (Tsao, 2010). The majority of polyphenols in plants exist as glycosides with different sugar units and acylated sugars at different positions of the polyphenol skeletons. Different polyphenol subgroups may differ significantly in stability, bioavailability and physiological functions related to human health. Numerous recent studies illustrate their significance in the prevention of the degenerative diseases (cancer, cardiovascular and neurodegerative diseases etc.) related to oxidative stress caused by reactive oxygen and nitrogen species (Dashwood, 2007; Khan et al., 2013; Quiñones et al., 2013; Semalty et al., 2010; Tangney and Rasmussen, 2013). The strong antioxidant activity of polyphenols is already evidenced *in vitro* and *in vivo* and ascribed to neutralization of free radicals by donating an electron or a hydrogen atom thus suppressing the generation of free radicals and reducing the rate of oxidation by inhibiting the formation of or deactivating the active species and precursors of free radicals or by neutralizing the free radicals, therefore stopping the chain reactions (Tsao, 2010). Polyphenols are also known as metal chelators and co-antioxidants involved in the regeneration of essential vitamins (Zhou et al., 2005). Some other mechanisms of antioxidant activity are also proposed, including inhibition of xanthine oxidase (Disilvestro, 2001) and induction of glutathione peroxidase, catalase and superoxide dismutase (Du et al., 2007). Formulation of poorly soluble polyphenols is a difficult task. Bioavailability of such natural products is usually determined by an appropriate balance between their hydrophilicity (for dissolving into the biological fluids) and lipophilicity (to cross lipidic biomembranes) (Semalty et al., 2010). Bioavailability is one of the most important parameters in pharmacokinetics, directly indicating the amount of unchanged drug that reaches the systemic circulation. Inadequate bioavailability is a result of both limited absorption and the first-pass metabolism. Many polyphenols having good aqueous solubility exhibit poor systemic absorption because of their large size, incompatibility with a process of passive diffusion and/or their poor miscibility with oils and other lipids (Semalty et al., 2010). As a result, their ability to cross the lipid-rich outer membrane of small intestine enterocytes is severely limited. Absorption of polyphenols from the gastrointestinal tract is poor and the estimated concentrations in plasma are low (Halliwell, 2008; Manach and Donovan, 2004; Rechner et al., 2002; Williamson et al., 2005). Application on the skin has similar limitations, since the size of active molecules is essential in addition to adequate lipophilicity of the formulation. In order to overcome these limitations a novel approach in phytoconstituents delivery known as *phytosomes* or *herbosomes* has been developed and numerous commercially available ingredients for pharmaceutical and personal care usage are already available.

PHYSICOCHEMICAL CHARACTERISTICS AND APPLICATION OF PHYTOSOME ENCAPSULATED POLYPHENOLICS

Phytosomes are cell like structures which result from the self-assotiation of the complexes obtained by a stoichiometric reaction of the standardized extract or polyphenolic constituents (e.g., flavonoids, terpenoids, tannins, xanthones) with the phospholipids (like phosphatidylcholine, phosphatidylserine) (Singh et al., 2011). Typically, a weight ratio between polyphenol phytoconstituents and phospholipids ranges from 1:1.5 to 1:4. A complex formed may show significant differences depending on the type of solvent and phospholipids used in the preparation. Commercially available phytosomes are prepared in protic or aprotic solvents, and phospholipids such as phosphatidylcholine or phosphatidylserine are usually employed (Semalty et al., 2010). As a result, products with different characteristics are available. The first phytosome generation was prepared by combining selected polyphenols or polyphenol extracts with phospholipids in nonpolar solvents (Bombardelli et al., 1989), but, more recently, a newer generation was developed using hydro-ethanolic solvents (Semalty et al., 2010). The traditional approaches for preparing the phyto-phospholipid complexes using harmful organic solvents like tetrahydrofuran and dichloromethane have now been widely replaced by the hydrophilic solvents like ethanol, thus increasing their potential pharmaceutical and cosmetic applications (Khan et al., 2013; Semalty et al., 2010). The solvent evaporation technique has been frequently utilized as a conventional method for formulating the phyto-phospholipid complexes. However, this process involves a number of processing steps which is a time consuming task and the quality of the end product depends upon the method adopted for drying of the residue. Newer technique with supercritical fluids can overcome the drawbacks of conventional methods enabling better control of the particle size and its distribution, and is more appropriate for sensitive substances (Khan et al., 2013).

Physotomes are vesicles similar to liposomes, formed from amphiphilic complexes of active substance and phospholipids. In liposomes the hydrophilic active substances are dissolved in the central cavity, with no possibility of molecular interaction with the surrounding phospholipid bilayer. On the contrary, spectroscopic evaluation and thermal analysis of complexation and molecular interactions between phytoconstituents and phosphatidylcholine revealed that in the phytosome complex the polar functionalities of the active substances interact *via* hydrogen bonds with the polar head of phospholipids forming a unique arrangement different from the physical mixture of its two constituents (Djekic et al., 2015; Khan et al., 2013; Semalty et al., 2010). Chemical bond within its structure makes phytosomes more stable than liposomes without altering biodegradability of the complex. In addition, the amount of phospholipids in liposomes is almost five-fold higher than in phytosomes, which makes them unsuitable as carriers containing therapeutic dose of active ingredient, particularly for oral route of administration. In phytosomes the molar ratio of phospholipids and active principles is usually of 1:1 or 2:1, thus higher in comparison with liposomes and more promising for carrier applications. A difference in the size of these vesicles is also visible, namely, the size of liposomes is much bigger than that of phytosomes (Djekic et al., 2016; Li et al., 2015).

A numerous of commercially available pharmaceutical and cosmetic ingredients have been produced by applying patented Phytosome® technology developed by Indena (Milan, Italy) in the late Eighties, using the standardized herbal extracts. Phytosomes as carriers for active substances can be formulated as preparations for both oral (powders, capsules, tablets) and topical applications (solutions, suspensions, emulsions, syrups, lotions, gels, creams) (Khan et al., 2013). Dosage forms for parenteral administration have not been developed so far, but the possibility to use phyto-phospholipid complexes for liposomes design (i.e., to yield a new type of supramolecular aggregates) was investigated. The resulting associates called phyto-liposomes provide an extended and continuous release of the active principles and drug safety (Angelico et al., 2014).

BIOMEDICAL RELEVANCE OF PHYTOSOME ENCAPSULATED POLYPHENOLICS

The phytosome technology enables conversion of phytoconstituents into molecular complexes with phospholipids (Semalty et al., 2010). Naturally occurring phospholipids incorporate an unsaturated fatty acid (such as oleic acid, linoleic acid or arachidonic acid) in position 2 and a saturated one (such as stearic acid or palmitic acid) in position 1 (Changediya et al., 2011; Kidd, 2009). The most commonly used phospholipids are those derived from soybean containing higher proportions, that is about 76% of phosphatidylcholine, with a high content of polyunsaturated fatty acids like linoleic acid (about 70%), linolenic acid and oleic acid (Scholfield, 1981). Nevertheless, a high percentage of unsaturated fatty acids present in their structure, make them sensitive to oxidation. Other representatives of glycerophospholipids are phosphatidylserine, phosphatidylethanolamine, phosphatidylinositol, phosphatidylglycerol and cardiolipin. The fatty acids within their structures are palmitic, stearic and linoleic acid. The soy phospholipids are absorbed at an extent greater than 90% in humans and reach peak plasma concentration in about 6 h after oral administration (Khan et al., 2013). The soy phospholipids have been shown to be free from many acute or chronic effects on laboratory animals at amounts even higher than the recommended doses and have been found to be non-immunogenic too. No teratogenic and carcinogenic effects have been reported either (List, 2015). Furthermore, phospholipids are a decent source of phosphatidylcholine and choline, both of which liquefy the fat dumped inside the liver in case of hepatic steatosis or fatty liver and exhibit other hepatoprotective effects as well (Kapoor Silki et al., 2012). They have also been reported to aid in clearance of serum cholesterol and increase circulating high-density lipoproteins (HDL) levels in plasma (Cohn et al., 2008). The presence of proportionally larger amounts of poly-unsaturated fatty acids in soy phospholipids makes it potentially useful in reducing the risk of coronary heart disease (Khan et al., 2013). In the human body they serve as transport molecules and surfactants (Kidd, 2002; Li et al., 2015). The basic goal of the phytosome approach was to combine the properties of phosphatidylcholine as a basic building molecule of cell membranes and polyphenolic substances to obtain the amphiphilic complexes that are simultaneously mixed with water and lipids. Due to the presence of physiological phospholipids in their structure these carriers can easily penetrate through the phospholipid bilayer membrane. Compared to ingredients which are not combined with phospholipids in the complex, phitosome approach enables the improvement of polyphenolic substances bioavailability, without resorting to pharmacological adjuvants or structural modification. Furthermore, their *in vivo* stability is increased and accompanied with a slower rate of elimination (Khan et al., 2013; Semalty et al., 2010). The potential for enhancement of biological effects such as cardiovascular, antiinflammatory, hepatoprotective, anticancer, hypoglycemic, weight loss, and antiaging, of the selected phytosome-encapsulated polyphenolics (Table 1) is illustrated in detail within the further subsections.

Table 1. The selected marketed polyphenols/phospholipid complex based ingredients

Natural source	Phytoconstituents in complex	Marketed ingredients	Dose	Mechanism of action
Silybium marianum	Silybin, silymarin silidianin, silicristin, izosilibin from Thistle seed	Silybin Phytosome ® (Siliphos®) Silymarin Phytosome®	120-200 mg	Increases glutathione levels in the liver.
Panax ginseng	Ginsenosides from rhizomes of ginger	Ginseng Phytosome ® Ginselect® Phytosome®	150 mg	Increases the level of enzymes (CAT, SOD, GPx, GR).
Camellia sinensis	Ginsenosides from rhizomes of ginger Epigallocatechin, catechin, epicatechin 3-O-gallate, epigallocatechin-3-O-gallate of green tea leaf	Green tea Phytosome® Greenselect ® Phytosome®	400 mg	Inhibits urokinase (an enzyme responsible for the growth of the tumor). Increases the activity of the enzyme GPx, GR.
Ginkgo biloba	Ginkgo flavonoids, ginkgo acids from a ginkgo - flavonglukozida, ginkgolide and bilobalide from ginkgo leaf	Ginkgoselect Phytosome® Ginkgo biloba terpenes Phytosome® Ginkgo biloba dimeric flavonoids Phytosome ® Virtiva®	120 mg (1.5% for topical application)	It stimulates the release of catecholamines and inhibits MAO and SOMT. Dilates capillaries and arterioles. Ginkgolide inhibit binding RAF. Flavonoids inhibit phosphodiesterase cAMP.
Vitis vinifera	Resveratrol, quercetin, catechin, epicatechin, procyanidins from grape seed	Leucoselect Phytosome® Masquilier' s Phytosome ®	50 - 100 mg	Protects endothelial cells from damage caused by peroxynitrite. Stimulates release of NO. It prevents oxidation of LDL.
Curcuma longa	Curcuminoides from the rhizome of turmeric	Curcumin Phytosome® Curcuvet® Meriva®	250 - 360 mg	Inhibits the metabolism of arachidonic acid. COX, LOX, cytokines, TNF release of steroid hormones. Stabilizes the lysosomal membrane.

PHYTOSOME ENCAPSULATED POLYPHENOLICS WITH EFFECT ON THE CARDIOVASCULAR SYSTEM

The development of Phytosome® technology has broaden potential application of *Ginkgo biloba* (foliuim) and *Vitis vinifera* (seeds et fructus) extracts in the treatment of cardiovascular disorders (Semalty et al., 2010). Standardized ginkgo extracts are widely used for the treatment of cerebral insufficiency (i.e., a short-term memory impairment, confusion and cognitive disorder which is related to impaired cerebral circulation and aging) and disorders of the peripheral circulation (such as Raynaud syndrome and intermittent claudication) (De Feudis, 1991; Mahadevan and Park, 2008). They typically contain 5-7% terpene trilactones, 22-27% of flavonol glycosides, and less than 5 ppm of ginkgolic acids (http://buecher.heilpflanzen-welt.de/BGA-Commission-E-Monographs/0183.htm). Ginkgolides and bilobalide are terpenoides with potential neuroprotective, vasoprotective, antiischemic and antiedematous actions. Flavonoids are related with enhancement of memory *via* increasing concentration of catecholamines and biogenic amines which level drops during the aging. Commercially available phytosomes with phytoconstituents from *Ginkgo biloba* leaf are Ginkgoselect® Phytosome® (phosphatidylcholine complex) and Virtiva® (phosphatidylserine complex). Virtiva® being active at lower doses than the standard phosphatidylcholine complex, since it was evidenced that phosphatidylserine was important for functioning of the central nervous system (Kennedy et al., 2007; Morazzoni et al., 2005). The efficacy of Ginkgoselect® Phytosome® to treat peripheral vascular disease was found to be 30 – 60% higher than that of Ginkgoselect® (Muir et al., 2002). The cardioprotective effects of Ginkgoselect® Phytosome® based on antioxidative activity was also reported (Panda and Naik, 2008). In this study, levels of marker enzymes (aspartate transaminase (AST), *lactate dehydrogenase* (LDH), and creatine phosphokinase (CPK)) were assessed in serum and heart, while antioxidant parameters viz., reduced glutathione (GSH), superoxide dismutase (SOD), catalase (CAT), glutathione peroxidase (GPx) and glutathione reductase (GR) and malondialdehde (MDA), were assayed in heart homogenate after isoproterenol-induced cardiotoxicity. In addition, histopathological examinations were performed. All results confirm the cardioprotective effect of Ginkgoselect® Phytosome®.

The total extractable phenolics in grapes are present at ≤ 10% in pulp, 60 – 70% in the seeds and 28 – 35% in the skin. The most abundant phenolics isolated from grape seeds and skins are flavan-3-ols (catechin and epicatechin) and their oligomers and polymers (proanthocyanidins) (Shi et al., 2003). Grape polyphenols (anthocyanins, flavanols, flavonols and resveratrol) possess many biological activities, such as antioxidant, cardioprotective, anticancer, antiinflammation, antiaging and antimicrobial (Xia et al., 2010). They have been known for a significant improvement of *low-density lipoprotein* (LDL) resistance to oxidation and their capillary protective effects ascibed with biological activity of catechin, epicatechin and oligomeric proanthocyanidins. Leucoselect® Phytosome® is commercially available complex of grape seed polyphenols. Its cardiovascular protecting activity is supported by clinical trials (Natella et al., 2002; Nuttal et al., 1998; Vigna et al., 2003). Leucoselect® Phytosome® was administered for 5 days to 20 young subjects in a single-blind randomized placebo-controlled crossover trial. A significant increase of serum total antioxidant capacity was observed from 30 min postdose with a further increase at 60 min postdose, in comparison with baseline values (Nuttal et al., 1998). The capacity in prevention of the plasma oxidative stress after a fatty meal rich in lipidic peroxides, has been evaluated in 8 healthy volunteers. At the beginning of the trial the subjects received the lipidic peroxides rich meal and after a week the same meal and Leucoselect® Phytosome®. The oxidative stress induced by the meal with the investigated phytosome product was decreased i.e., the plasma postprandial lipid hydroperoxide concentration was significantly reduced with an increase of *total reactive antioxidant potential* (TRAP) and resistance of LDLs to oxidative modification (Natella et al., 2002). The potential to decrease low-density lipoprotein susceptibility to oxidation and oxidative stress damage was investigated in a group of heavy smokers (Vigna et al., 2003). Leucoselect® Phytosome® was administered for 4 weeks to 24 healthy male heavy smokers, aged ≥ 50, in a randomized double-blind crossover trial. A significant improvement of LDL resistance to oxidation was induced, as shown by lipid peroxidation parameters (thiobarbituric acid reactive substances concentration (TBARS)), an index of lipid peroxidation and oxidative stress was significantly reduced, while the lag phase (an index of LDL resistance to oxidation) was prolonged, both in comparison with placebo and basal values (Vigna et al., 2003). In onother study (http://www.indena.com/pdf/leucoselect_phytosome_int.pdf), Leucoselect® Phytosome® was administered for 4 weeks to 24 subjects with sugar metabolism challenges in a double blind crossover parallel study, significantly reducing urinary excretion of 8-epi-PGF2α (a marker of oxidative stress) in comparison with placebo.

PHYTOSOME ENCAPSULATED POLYPHENOLICS WITH ANTIINFLAMMATORY ACTIVITY

Milk thistle (*Silybum marianum* (L.) Gaertn) is known as an herbal drug with significant anti-inflammatory activity. Positive effect of *Silybum marianum* extract on ulcer healing, cirrhosis and chronic inflammatory diseases was evidenced after their oral application. Silymarin, a flavonolignan from milk thistle is a complex mixture of four flavonolignan isomers, namely silybin, isosilybin, silydianin and silychristin (Figure 1).

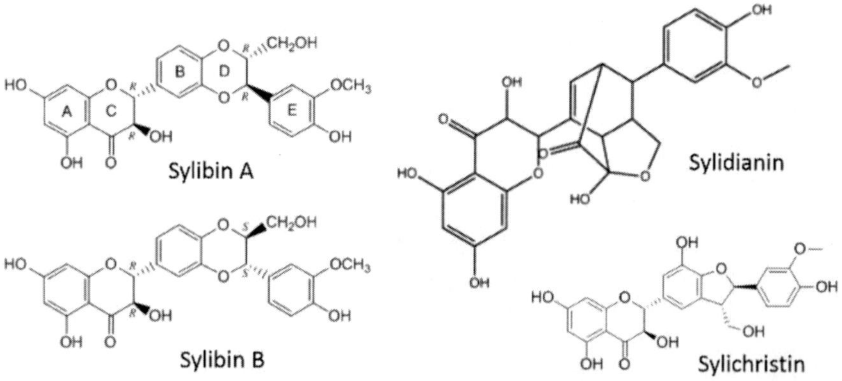

Figure 1. The chemical structure of the main sylimarin constituents.

These active constituents can be distinguished on the basis of covalent interaction between its flavonol and lignan constituents. These flavonolignans were found to have free radical scavenging and antiinflammatory properties (Pradhan and Girish, 2006; Tsao, 2010). Antiinflammatory activity was also noticed after topical application on the skin (Katiyar, 2005). To evaluate the anti-inflammatory activity, silymarin phytosomes were applied topically as microdispersion in water and the reduction of edema was used as the end point. Croton oil induced dermatitis was used as an inflammation model involving both tissue and vascular damage. Phospholipids are *per se* devoid of anti-inflammatory activity, but it was found that their complexation with silymarin prolonged its activity. The silymarin complex showed greater reduction of edema (76% reduction in 6 h) as compared to that with the free form (33% even after 12 h) in the croton oil test in mice. Therefore, the phytosome was about two-fold more effective than the free phytoconstituent (Bombardelli 1994).

According with the manufacturer's claims Silymarin Phytosome® in concentrations of up to 3% exhibits enhanced functionality of the flavanolignans resulting in a longer lasting soothing and antioxidant protectant activity and increased effectiveness on the skin, compared to the free active substances, due to a higher affinity of the complex for skin phospholipids. The free radical scavenging capacity of the extract was established *in vitro* by totally suppressing the release of reactive oxygen species by stimulated leukocytes, while in humans, Silymarin Phytosome® reduced the incidence of UVB induced erythema (http://www.indena.com/pdf/skin_care.pdf).

A standardized extract of *Boswellia serrata* gum resin, a traditional ayurvedic medicine known for its anti-inflammatory activity, is becoming increasingly interesting in modern therapy (Hüsch et al., 2013; Siddiqui, 2011). The resinous part of *Boswellia serrata* possesses monoterpenes, diterpenes, triterpenes, tetracyclic triterpenic acids and four major pentacyclic triterpenic acids (i.e., β-boswellic acid, acetyl-β-boswellic acid, 11-keto-β-boswellic acid and acetyl-11-keto-β-boswellic acid) (Figure 2), responsible for inhibition of pro-inflammatory enzymes. Acetyl-11-keto-β-boswellic acid is the most potent inhibitor of 5-lipoxygenase, an enzyme responsible for inflammation (Siddiqui, 2011).

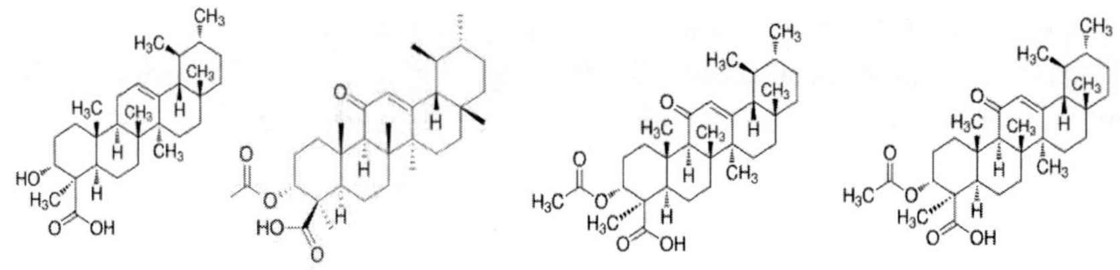

β-boswellic acid   Acetyl-β-boswellic acid   11-keto-β-boswellic acid   Acetyl-11-keto-β-boswellic acid

Figure 2. The chemical structure of the main sylimarin constituents.

A potential for *Boswellia serrata* gum resin extracts to alleviate a variety of inflammatory conditions that include inflammatory bowel disease, rheumatoid arthritis, osteoarthritis and asthma were established in numerous *in vitro* assays, animal studies and pilot clinical trials (Abdel Tawab et al., 2011). Since 2002 *Boswellia serrata* gum resin extract has the status of the *orphan drug* for the treatment of peritumoral brain edema in Europe (Skarke et al., 2012). The mechanisms underlying the anti-inflammatory action of *Boswellia serrata* gum resin active ingredients are under research. The investigations in this field suggested that two quantitatively minor boswellic acids (11-keto-β-boswellic acid and acetyl-11-keto-β-boswellic acid) could interfere with the production of leukotrienes by inhibition of 5-lipoxygenase, while other boswellic acids, particularly β-boswellic acid, could targeting the microsomal prostaglandin (PG) E2 synthase-1 (mPGES-1) as well as cathepsin G (CatG) (Abdel Tawab et al., 2011; Siemoneit et al., 2012; Skarke 2012; Tausch et al., 2009). Phytosomal complex with enhanced bioavailability is Casperome® (Boswellia Phytosome®) formulated with soybean phospholipids (Hüsch 2013; http://www.indena.com/casperome/). In a functional study on Casperome® as complementary intervention in asthmatic patients, the subjects have been randomized to receive the phytosome (500 mg/day) or no additional treatment, for a period of 4 weeks. It was shown that Casperome® can help reduce the need for inhaled corticosteroid and beta-agonist up to 43% in subjects with asthma (Ferrara et al., 2015). The concentration of the six major boswellic acids (11-keto-β-boswellic acid, acetyl-11-keto-β-boswellic acid, β-boswellic acid, acetyl-β-boswellic acid, α-boswellic acid, and acetyl-α-boswellic acid) was evaluated in rats (the plasma and in a series of tissues) (Hüsch et al., 2013). These ingredients are pentacyclic triterpenic structures. Their concentrations were determined in highly vascularized tissues (brain, liver, and kidney) and low vascularized tissues (muscles and eyes). Results indicated that the plasma concentrations of these acids were several times higher after application of this product compared to those after application of the extract. More significant concentration increase was in the tissues (from 17 times higher concentration in the eye and up to 35 times higher concentrations in the brain) (Hüsch et al., 2013).

PHYTOSOME ENCAPSULATED POLYPHENOLICS WITH HEPATOPROTECTIVE ACTIVITY

In addition to previously described anti-inflammatory effects of milk thistle ingredients, the fruit of the plant, has been a liver support remedy for 2000 years (Kidd, 2009). The most important ingredient of the fruit is also a silymarin complex. Silymarin has been shown to be effective in the treatment of liver diseases of various kinds, including hepatitis, cirrhosis, fatty infiltration of the liver (chemical and alcohol induced fatty liver), and inflammation of the bile duct (Cacciapuoti et al., 2013; Féher and Lengyel, 2012; Ferenci et al., 2016; Fried et al., 2012; Saller et al., 2001). The antioxidant properties of silymarin substantially improve the liver resistance to damage caused by toxic agents such as acetaminophen, alcohol, carbon tetrachloride, and the mushroom toxins (Kidd, 2009). Silybin, the main and most potent ingredient of the silymarin complex, occurs as a mixture of two diastereoisomers (silybin A and silybin B). Activity of phytosomes is based on glutathione conservation in the parenchymal cells, while phosphatidylcholine helps repair and replace cell membranes. In this way, these constituents act synergistically, sparing liver cells from destruction (El-Lakkany et al., 2012; Valenzuela et al., 1989). Silybin, as well as silymarin, have a poor intestinal absorption. Therefore, many studies have evaluated soy phospholipid-complexed ingredients such as Silipide® or the pharmacokinetically equivalent Siliphos®, and demonstrated the enhanced absorption of the complex. Morazzoni et al., (1993) evaluated the plasma level profile and the biliary excretion of silybin in rats, after single equimolar oral doses (200 mg/kg, expressed as silybin equivalents) of Silipide® and of silymarin. Silipide® silybin reached peak plasma levels within 2 h, with a C_{max} of 9.0 ± 3.0 µg/ml for unconjugated drug and 93.4 ± 16.7 µg/ml for total (free + unconjugated) drug. Maximum total biliary concentrations of silybin (2989 ± 568 µg/ml) were observed within 2 h and the biliary recovery after 24 h accounted for about 13% of the administered amount. After administration of silymarin, unconjugated and total plasma silybin levels as well as biliary excretion were several-fold lower than those observed after treatment with silipide. Silybin recovered over a 24 h period after silymarin intake accounted for about 2% of the administered dose. Plasma and bile obtained after administration of silymarin contained also silydianin, silycristin and, to a greater extent, isosilybin. The concentrations of the latter compound in plasma and in bile were higher than those of silybin itself. The relative bioavailability of Silipide® (calculated in the target organ as the ratio between AUCs of cumulative biliary excretion curves) was 10-fold higher than that of silymarin. In another study on rats, silybin given as Silipide® at 200 mg/kg was detectable in the plasma within minutes, peaked after 1 h, and its plasma levels remained elevated for over 6 hours. The AUC value of free silybin after Silipide® intake was more than 400% higher than after intake of silymarin and is rapidly reach the liver and appear in the bile within 2 h at a level at least 6.5 times higher than that from non-complexed silybin administered as silymarin (http://www. siliphos.info/scientific-support/#). After oral administration in rodents in various models of liver damage, Silipide® exhibited a significant and dose related protective effect against hepatotoxicity induced by chloroform, acetaminophen, ethanol and galactosamine (Conti et al., 1992). The current drugs used for this purpose have a limited and a variable hepatoprotective effect, and on the other hand, exhibit significant adverse effects. In a randomized study, patients administered 240 mg of Silipide® or placebo twice daily for a period of three months. The Silipide® group showed significant lowering of both serum alanine transaminase (ALT) and AST, while in the placebo group these markers worsened. The Silipide® treatment was well tolerated, with fewer adverse events than for the placebo group, and no patient discontinued the trial due to adverse effects. Side effects that were reported included nausea, dyspepsia, heartburn and transient headache in 5.2% of patients compared to 5.1% in the placebo group. Uncomplexed silybin was worst tolerated with adverse effects in 8.2% of patients (Marena and Lampertico, 1991). The pharmacokinetic studies of Silipide® in healthy human subjects showed the 4.6 fold increased oral bioavailability of silybin compared with silymarin (Barzaghi et al., 1990). In a pilot study on the liver protective effect of Silipide® in chronic active hepatitis it was found the improved liver function tests related to hepatocellular necrosis and/or increased membrane permeability in patients affected by chronic active hepatitis (Buzzelli et al., 1993). The iron chelating ability of Silipide® was demonstrated in patients with chronic hepatitis C and Batts–Ludwig fibrosis, associated with a reduction in iron stores in patients with advanced fibrosis (Bares et al., 2008). The effect of a silybin–vitamin E–phospholipid complex on nonalcoholic fatty liver disease was investigated in a pilot study (Loguercio et al., 2007). Among indexes of fibrosis, body mass, degree of steatosis, insulinemia, plasma levels of TGF-β, TNFα, and γ-GTP, significant correlation was observed. Reduction in ultrasonographic scores for liver steatosis and improvement of liver enzyme levels, hyperinsulinemia and indexes of liver fibrosis suggest that this complex could be used as a complementary approach to the treatment of patients with chronic liver damage. Also, it was determined that Silymarin Phytosome® showed better antihepatotoxic activity than silymarin alone against the toxic effects of aflatoxin B1 on broiler chickens (Tedesco et al., 2004). Several studies on silymarin administered as Silymarin Phytosome® found that the complex could protect the fetus from maternally ingested ethanol (La Grange et al., 1999). Additionally, in another study it was also reported that a better fetoprotectant activity against ethanol-induced behavioral deficits was obtained with Silymarin Phytosome® than with uncomplexed silymarin on fetuses of both sexes (Busby et al., 2002).

Ginkgoselect Phytosome® proved to be effective in treatment of liver damage after administration of antituberculosis drug rifampicin. This effect was related to its antioxidant activity, with in addition to free radical scavenging acted indirectly by stimulating the natural protection system. As a result, levels of glutathione and catalase, superoxide dismutase, glutathione peroxidase, and glutathione reductase were increased. The net effect was inhibition of lipid peroxidation product and production of malondialdehyde which is responsible for the damage to cell membranes and consequently, the liver tissue. The levels of these markers of oxidative stress were determined in liver homogenates, in addition to determination of serum transaminase levels, alkaline phosphatase, albumin and total proteins as indicators of general liver functions (Naik and Panda, 2008).

More of natural substances (herbal drugs) were investigated as hepatoprotectants and their complexes with phospholipids have been developed. Some of these substances flavonoids that act as antioxidants, such as quercetin, curcumin and naringenin (Figure 3).

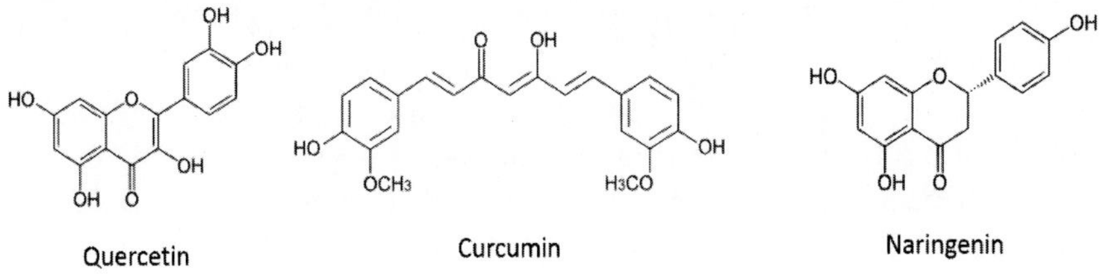

Quercetin   Curcumin   Naringenin

Figure 3. The chemical structure of the selected flavonoids with biomedical relevance, available also in phytosome form.

Traditionally, they have been used in treatment of conditions such as chemical and alcohol induced fatty liver. A quercetin–phospholipid complex was also developed, and investigated in a murine model of liver injury (carbon tetrachloride-induced damage). Markers level for oxidation in liver homogenate were evaluated during the treatment with 10 mg / kg and 20 mg / kg of free quercetin and quercetin / phospholipid complexes. Complexed form of quercetin showed significant antioxidant activity (Maiti et al., 2005).

PHYTOSOME ENCAPSULATED POLYPHENOLICS WITH ANTICANCER PROPERTIES

Further researches on silymarin and silybin activity revealed their anticancer and chemopreventive properties. The antiradical potential and the cytoprotective activity may be responsible for the use of silymarin as a chemoprotective and anticancer agent. By inhibiting the carcinogenic potential of a numerous substances (Dorai and Aggarwal, 2004), they have the potential to be successfully used in the adjuvant therapy of cancer. Besides prevention of tissue damage caused by oxidative stress of various chemotherapeutics, these products prevent hepatotoxicity, a major side-effect of many chemotherapy treatments. Silybin and its phospholipid complex (IdB1016) have been shown to potentiate the effect of cisplatin in mice bearing human ovarian cancer cells (Giacomelli 2002). After repeated application of the complex, the growth of tumor cells was significantly reduced (Gallo et al., 2003). The modifying effect of dietary administration of silymarinon azoxymethane (AOM)-induced colon carcinogenesis, being putative precursor lesions for colonic adenocarcinoma, was investigated in male F344 rats in the short-term as well as the long-term studies. The activity of detoxifying enzymes (glutathione S-transferase (GST) and quinone reductase (QR)) in liver and colonic mucosa was determined in rats gavaged with silymarin. Subsequently, the possible inhibitory effects of dietary feeding of silymarin on AOM-induced colon carcinogenesis were evaluated using a long-term animal experiment. In the short-term study, dietary administration of silymarin (100, 500 and 1000 ppm in diet), either during or after carcinogen exposure for 4 weeks, caused significant reduction in the frequency of colonic aberrant crypt foci (ACF) in a dose-dependent manner. Silymarin given by gavage elevated the activity of detoxifying enzymes in both organs. In the long-term experiment, dietary feeding of silymarin (100 and 500 ppm) during the initiation or postinitiation phase of AOM-induced colon carcinogenesis, reduced the incidence and multiplicity of colonic adenocarcinoma. The inhibition by feeding with 500 ppm silymarin was significant ($p < 0.05$ by initiation feeding and $p < 0.01$ by postinitiation feeding). Also, silymarin administration in the diet lowered the proliferating cell nuclear antigen (PCNA) labeling index and increased the number of apoptotic cells in adenocarcinoma. Beta-glucuronidase activity, PGE(2) level and polyamine content were decreased in colonic mucosa.

Curcumin, a hydrophobic polyphenol derived from the rhizome of the turmeric plant (*Curcuma longa*, family Zingiberaceae) also have potential for application in this area. Curcumin is derived from the spice turmeric and has anti-inflammatory and antineoplastic effects *in vitro* and in animal models, including preventing ACF and adenomas in murine models of colorectal carcinogenesis. Inhibiting the production of the procarcinogenic eicosanoids prostaglandin E2 (PGE2) and 5-hydroxyeicosatetraenoic acid (5-HETE) can suppress carcinogenesis in rodents. Curcumin reduces mucosal concentrations of PGE2 (*via* inhibition of cyclooxygenases 1 and 2) and 5-HETE (*via* inhibition of 5-lipoxygenase) in rats (Carroll et al., 2011). Curcumin blocks the cell transformation, proliferation and invasion of tumor tissue, both *in vitro* and *in vivo*. Also, it induces cell apoptosis of cancer cells in culture, while has no influence on the normal cells (Kidd, 2009; Zhang et al., 2013). Alone and in combination with other substances is effective in multiple myeloma, osteosarcoma, colorectal cancer, pancreatic cancer and prostate cancer. This activity is realized primarily by suppression of factor NF-κB transcription, and late on of pSTAT-3 and AP-1 and VEGF growth factors (Kidd, 2009; Siviero et al., 2015). These factors are essential for the expression of genes involved in processes such as proliferation, angiogenesis, metastasis, and resistance to chemotherapy. There are currently a number of clinical trials in different phases with the aim to investigate the curcumin alone or in combination with other therapeutic agents in a variety of cancers. The focus is on chemoprevention in patients with pre-malignant lesions, slowing the progression of the disease that has already developed and moderate resistance to existing therapy (Singhal et al., 2009). Caroll et al. (2011) assessed the effects of oral curcumin (2 g or 4 g per day for 30 days) on PGE2 within ACF (primary endpoint), 5-HETE, ACF number, and proliferation in a nonrandomized, open-label clinical trial in 44 eligible smokers with eight or more ACF on screening colonoscopy. Forty-one subjects completed the study. Neither dose of curcumin reduced PGE2 or 5-HETE within ACF or normal mucosa or reduced Ki-67 in normal mucosa. A significant 40% reduction in ACF number occurred with the 4-g dose (P < 0.005), whereas ACF were not reduced in the 2-g group. The ACF reduction in the 4-g group was associated with a significant, five-fold increase in posttreatment plasma curcumin/ conjugate levels (versus pretreatment; P = 0.009). Curcumin was well tolerated at both 2 g and 4 g. In phase I clinical studies (Hsu and Cheng, 2007) curcumin with doses up to 3600-8000 mg daily for 4 months did not result in discernible toxicities except mild nausea and diarrhea. Although the pharmacologically active concentration of curcumin could be achieved in colorectal tissue in patients taking curcumin orally and might also be achievable in tissues such as skin and oral mucosa, which are directly exposed to the drugs applied locally or topically, the pharmacokinetic studies of curcumin indicated in general a low bioavailability of curcumin following oral application. Meriva® is a phytosomal form of curcumin, formulated with soy lecithin in a 1:2 weight ratio with an overall content of the active substance of around 20%. Curcumin is sparingly soluble substance both in water and in oily solvents, but shows polar groups (two phenolic hydroxyl and one enolic hydroxyl) that can interact *via* hydrogen bondings and polar interactions with complementary groups, like the polar heads of phospholipids. The complex shows a high affinity for biological membranes and enhanced cellular captation (Kidd, 2009; Semalty et al., 2010). The maximum plasma concentrations of curcumin in rats was 5 times higher after administration of phytosomes complex compared to curcumin alone. In a comparative pharmacokinetic study in humans (Cuomo et al., 2011), the absorption of each single curcuminoid present in commercial curumin was compared between two dosages of Meriva® (1.0 g and 1.9 g, corresponding to 209 and 376 mg of curcuminoids, respectively) and one dosage of the corresponding unformulated curcuminoid mixture (1.8 g). The overall increase of curcuminoid absorption from Meriva® was 27-fold for the low dosage and 31-fold for the high dose (Anand et al, 2007; Giori and Franceschi 2007). The improved oral bioavailability of curcumin as Meriva® has been translated into clinical efficacy for addressing the natural inflammatory response function at dosages significantly lower than those associated to unformulated curcumin, with ongoing clinical studies also for other conditions where a solid mechanistic rationale and preclinical rationale and preclinical evidence of efficacy exists for curcumin. Apart from animal data (LD50 >2 g/kg in rats), no side effects were observed when Meriva® was administered at 1.2 g/day to over 100 volunteers for 18 months (Belcaro et al., 2010). The improvement of oral bioavailability of curcumin, is expected to significantly improve therapy in the future.

PHYTOSOME ENCAPSULATED POLYPHENOLICS WITH HYPOGLYCEMIC PROPERTIES

Berberine (BER) is isoquinoline alkaloid which has caused a growing interest in recent years due to its antidiabetic effects. It occurs naturally in numerous natural medicinal plants such as in the Ranunculaceae and Berberidaceae families. It is a potential natural alternative to other synthetic antidiabetic drugs. The gastrointestinal absorption of BER is poor and absolute bioavailability was reported to be no more than 1% (Godugu et al., 2014), since it is a substrate of P-gp. Thus, high dose is necessary in clinical conditions, which can trigger adverse side effects. The phytosomes loaded with berberine (P-BER) were administered to animals (Wistar rats). The animals, randomly picked and put into the control group and the other three groups, orally intragastric gavaged with BER (100 mg/kg, suspended in 0.5%

carboxymethyl cellulose sodium), P-BER (100 mg/kg), or RP-BER (100 mg/kg) for 4 weeks, respectively. The results showed that the P-and RP-BER demonstrated the significantly enhanced antidiabetic efficacy compared with the free BER by reducing the blood glucose levels to normal, probably due to an increase in insulin sensitivity followed with stimulating glucose utilization. Moreover, P-BER, RP-BER and BER showed similar trends in activity which were in agreement with the results of *in vitro* and *in vivo* studies. The increased triglycerides levels in the liver were lowered to near control ones after the treatment with P-BER and RP-BER for 4 weeks, while BER did not produce the same effect. New challenges in the treatment of diabetes are simultaneous control of hyperglycemia and hyperlipidemia due to the reduction of risk for atherosclerosis and myocardial infarction, as it was obtained with this complex in the very recent study (Yu et al., 2016). The phytosomes loaded with berberine-phospholipid complex (P-BER) were prepared by a rapid solvent evaporation method followed by a self-assembly technique. The P-BER showed a nanoscale particle size, a negative surface charge, and high drug entrapment efficiency (~85%). The oral bioavailability of the P-BER was significantly improved by 3-fold in comparison with the previous pharmacokinetic studies. The oral administration of P-BER could suppress the fasting glucose levels and improve the ability of systematic hyperlipidemia metabolism of db/db diabetic mice (Yu et al., 2016).

In addition to BER, silymarin also has proven to be effective in tests on rats with hereditary hypertriglyceridemia and insulin resistance independent of obesity (Oliyarnyk et al., 2014).

THE EFFECT OF PHYTOSOME ENCAPSULATED GREEN TEA POLYPHENOLS ON WEIGHT LOSS

Camellia sinensis L. has been traditionally used for making beverages (infusion drink). Tea leaf extract is a rich variety of ingredients with different effects. Polyphenols, including phenolic acids, such as gallic acid, are constituting up to 30% of the dry weight of tea leaves (Graham, 1992). Green tea infusion contains flavanols (catechin polyphenols), flavanols and phenolic acids. Flavanols from green tea, such as (-)-epigallocatechin 3-O-gallate and (-)-epigallocatechin, exhibited a marked antioxidant activity. In addition to the antioxidative activity of polyphenols, a beneficial effect on lipid profile and on weight loss have been observed lately. Such activity is attributed to derivatives of gallocatechin, particularly epigallocatechin 3-O-gallate as the most common ingredient. Other active ingredients are epigallocatechin, catechin, and epicatechin-3-O-gallate (Kidd, 2009). Oral bioavailability of flavonoids, both in aglyconic or glycosidic form, is low and erratic due to limited absorption, elevated presistemic metabolism and rapid elimination (Manach et al., 2005). Fort the purpose to increase the bioavailability of green tea polyphenols Greenselect® Phytosome® was produced from a standardized decaffeinated green tea extract (Greenselect®) and soy phospholipids in a 1:2 ratio, and contains 19-25% of catechins with the main constituent (-)-epigallocatechin 3-O-gallate at least 13%. Phytosome form has been proven to be more bioavailable compared to the unformulated extract. Twelve healthy male volunteers were randomly divided in two groups. One received a single dose of Greenselect® (containing 240 mg of tea catechins by HPLC). The second group received 1200 mg of Greenselect® Phytosome® (containing 240 mg of tea catechins by HPLC). Epigallocatechin 3-O-gallate was chosen as the biomarker for absorption and the peak concentration at 2 hours was more than doubled with phytosome product in comparison with to the nonformulated standardized decaffeinated green tea extract, while its plasma levels remain considerably higher (Pietta, 1998). For the assessment of a weight loss promotion effect a multicenter study was conducted, where all participants were on a low calorie diet and in addition, half of them received Greenselect® Phytosome®. During controlled diet, average energy intake was 1250-1350 kcal for women and 1650-1750 kcal for men on a daily basis, divided into at least four meals. The evaluated parameters were body weight, body mass index, waist circumference, total cholesterol, total triglyceride, and fasting blood glucose concentration at the beginning, after 45 days and after 90 days of the study period. Average weight loss was 6 kg in the placebo group and 14 kg in the group which received Greenselect® Phytosome®. This complex also exhibited a mild hypoglycemic activity by stimulating the endogenous insulin production. Further researches in this area are necessary.

WOUND HEALING EFFECT OF PHYTOSOME ENCAPSULATED SINIGRIN

Sinigrin is one of the glucosides that is naturally found in plants of the Brassicaceae family, such as Brussels sprouts, broccoli and black mustard seeds (Brassica nigra). The enzyme myrosinase breaks down glucosides to thiocyanate, isothiocyanate, and cyanide, which are the main bioactives known for biological activity (Chew, 1988). Besides its different biological activities, such as anticancer (Lazarus et al., 1994; Rungapamestry et al., 2008) antimicrobial (Shofran et al., 1998) and anti-inflammatory (Lee et al., 2014), the wound healing effect of sinigrin has also been investigated lately (Mazumder et al., 2016a). Wound healing is a complex process involving several inter-related biological and molecular activities for achieving tissue regeneration (Boateng and Catanzano, 2015). The main physiological events during this process include hemostasis, inflammation, proliferation and maturation. Healed wound is one in which the tissue again has a normal anatomic structure and function (Boateng et al., 2008). On the basis of the nature of the repair process, wounds can be classified as acute or chronic (Whittam, 2016). The main causes of acute wounds involve mechanical injuries such as abrasions and torn skin, penetrating wounds caused by knives or as a result of surgery. Acute wounds are generally tissue injuries which heal entirely, with little scarring and within a relatively short time period (generally 8 to 12 weeks) (Percival, 2002). The therapeutic effect of sinigrin–phytosome complex was evaluated *in vitro* on human keratinocyte cells cultivated in appropriate media. The wound area was formed artificially and visually observed after 15 h, 24 h and 42 h. An inverted microscope fitted with a camera was utilised for the cell images. Untreated cells and cells treated with a particular concentration of sinigrin and sinigrin–phytosome complex were monitored. The results clearly show that phytosome complex exhibited more significant wound healing effects compared to other wound treatment options. Very recently, sinigrin and its phytosome formulations have been investigated for its wound-healing actions (Mazumder, 2016b). The *in vitro* drug release, determined by using dialysis sacks, indicated a controlled and sustained release of sinigrin from the phytosome complex. *In vitro* Franz cell diffusion studies were performed on human abdominal skin. Tape stripping results showed that the sinigrin-phytosome complex (0.5155 µg/ml) statistically significantly enhanced the delivery of sinigrin into the stratum corneum-epidermis when compared to the free sinigrin (0.0730 µg/ml). These results are very affirmative regarding the utilization of sinigrin-phytosome complex for various skin-related diseases including wound healing.

ANTIAGING PROPERTIES OF
PHYTOSOME ENCAPSULATED POLYPHENOLS

Phytosomes technology offers interesting and novel applications for the use of polyphenolics as cosmetic active ingredients. Sericoside isolated from *Terminalia sericea*, the plant that has been traditionally used in Africa and Asia for the treatment of severe skin diseases, in the phytosome form has been evaluated and skin restructuring, capillary protecting activity, wound healing and antiedema properties were demonstrated (Bombardelli et al., 1986), due to a reduction of capillary permeability. The clinical study was carried out on 10 female Caucasian subjects (40 - 55 years), with crow's feet wrinkles radiating from the outer corner of the eye (Cristoni, 2005). Sericoside Phytosome® 0.5% emulsion and corresponding control formulation were applied twice-daily for 42 days on all crow's feet area, reproducing normal conditions of use. The product showed to be effective in over 70% of the tested subjects, accomplishing a reduction in wrinkles depth and total wrinkled surface. Compared to the blank product, the Sericoside Phytosome® containing emulsion has significantly increased the microrelief furrows (61%, p=0.035) and has shown to shift the skin wrinkles class distribution from deep to median and superficial.

Evaluation of Silymarin Phytosome® revealed an antioxidant, UV protectant and free radical scavenging activity. UV-induced erythema was assessed in 18 volunteers and it was reduced for 24% when the subjects were treated with a 5% Silymarin Phytosome® containing gel applied daily for seven days on two different areas, as compared to blank preparation. At the end of the treatment on both areas the Minimum Erythematous Dose (MED) has been determined (Singh et al., 2009). Chemiluminescence (CL) or emission of light by stimulated macrophages was used as a measure of macrophage activation and an indicator of the production and release of toxic oxygen species (including superoxide anion, hydroxyl radical, hydrogen peroxide) by stimulated leukocytes. Incubf macrophages (1 million/vial) with 1 mg of Silymarin Phytosome® was shown to produce total abrogation of macrophage CL with emission of light equivalent to background control levels (http://www.indena.com/pdf/silymarin_silymarinPhytosome_st_pc.pdf).

New evidence studies have underlined the properties of Siliphos® in providing an activity similar to retinoic acid, but devoid of any irritation potential retinoic acid may cause (Kitajima et al., 2009). Reduces the expression of keratinocytes terminal differentiation markers, thus providing a visible antiaging effect. It also reduces the expression of inflammatory response markers, modulating the Nf-KB (Nuclear Transcription Factor) expression, being devoid of any irritation potential (Kitajima et al., 2009). Furthermore, it was claimed that this ingredient stimulates the basement membrane protein expression, as laminin-5 and integrin β4, involved in the organization of the extracellular matrix, thus improving the tissue compactness. In a 3D skin model, Siliphos® stimulates the basal membrane and derma (Myiata et al., 2005), and protect from UV induced damage by reduction of Caspase-3 cleavage on HaCaT cells.

Extracts of *Panax ginseng* C.A. Meyer have been systematically used for the prevention or treatment of a variety of conditions frequently associated with aging. Ginselect® Phytosome® has been tested on a total of 60 healthy subjects (17 - 88 years) who were divided into different groups depending on the specific experimental procedure applied. The application of Ginselect® Phytosome® (ampoules of 10 mg/ml, applied daily) showed to ameliorate the hydration of the corneous layers with statistical significance. These findings have been confirmed in two additional separated trials conducted on 20 subjects each (treated for 30 days), that showed statistically significant amelioration of various dermatological and cosmetic parameters, such as hydration, trophism, elasticity or dryness of the skin of the face. The hydration of the *stratum corneum* was related to the liposomal-like properties of the phospholipids and a transdermic action of the ginseng saponins likely related to increased blood perfusion and nutrients delivery to the skin (http://www.indena.com/pdf/ginselect_phytosome.pdf), as it was confirmed by the regional increase of cutaneous temperature of the hemiface after application of Ginselect® Phytosome® in female subjects older than 40 years (Bombardelli et al., 1989).

CONCLUSION

Low bioavailability of polyphenolics of herbal origin, that have long been known for its protective and therapeutic properties, is the important factor that limits their biomedical exploitation. The complexes of such polyphenolic active substances and phospholipids, which form vesicles of the phytosome type, are of increasing importance in current pharmacotherapy and their potential benefits are already recognized. The efficiency of such novel ingredients is improved upon their comlexation and encapsulation in phytosomal vesicles. This new strategy has the potential to enable the development of various pharmaceutical products that can be used in the treatment of liver diseases, cardiovascular diseases, cancer, inflammatory and metabolic diseases, as well as in the local treatment of damage or skin diseases or in new cosmetic products. Until now, *in vitro* and *in vivo* studies (in animals and in humans) of the most of commercially available phytosomes were performed. Evaluation of efficacy and safety of a larger number of formulations is in various stages of clinical trials. These investigations look promising but more clinical trials are necessary to thoroughly characterize the potential to treat serious diseases in humans or for development of cosmetic products with significantly enhanced efficacy. For a complete understanding of the potential for the delivery of polyphenol active substances, it is necessary to carry out detailed examination of the stability of the complex itself, as well as the final product. Particular attention should be paid to the development and optimization of the production processes in order to achieve the desired safety, stability, and efficacy of the pharmaceutical and cosmetic products with phytosome encapsulated polyphenolics.

REFERENCES

Abdel Tawab, M., Werz, O., Schubert-Zsilavecz, M. (2011). Boswellia serrata — an overall assessment of in vitro, preclinical, pharmacokinetic and clinical data. *Clin. Pharmacokinet.,* 50, 349–69.

Anand, P., Kunnumakkara, A. B., Newman, R. A., Aggarwal, B.B. (2007). Bioavailability of curcumin: problems and promises. *Mol. Pharm.* 4, 807–18.

Angelico, R., Ceglie, A., Sacco, P., Colafemmina, G., Ripoli, M., Mangia, A. (2014). Phyto-liposomes as nanoshuttles for water-insoluble silybin–phospholipid complex. *Int. J. Pharm.,* 471, 173–81.

Bares, J. M., Berger, J., Nelson, J. E., Messner, D. J., Schildt, S., Standishm L. J., Kowdley, K. V. (2008). Silybin treatment is associated with reduction in serum ferritin in patients with chronic hepatitis C. *J. Clin. Gastroenterol.,* 42, 937–44.

Barnes, P. M., Powell-Griner, E., McFann, K., Nahin, R. L. (2002). Complementary and alternative medicine use among adults: United States. *Adv. Data,* 27, 1–19.

Barzaghi, N., Crema, F., Gatti, G., Pifferi, G., Perucca, E. (1990). Pharmacokinetic studies on IdB 1016, a silybin-phosphatidylcholine complex, in healthy human subjects. *Eur. J. Drug Metab. Pharmacokinet.,* 15, 333–8.

Belcaro, G., Cesarone, M. R., Dugall, M. Pellegrini, L., Ledda, A., Grossi, M. G., Togni, S., Appendino, G. (2010). *Altern. Med. Rev.,* 15, 337–44.

Boari, C., Montanari, F. M., Galletti, G. P., Rizzoli, D., Baldi, E., Caudarella, R., Gennari, P. (1981). Epatopatie tossiche professionali. Minerva Med., 72, 2679–88.

Boateng, J., Catanzano, O. (2015). Advanced Therapeutic Dressings for Effective Wound Healing-A Review. *J. Pharm. Sci.,* 104, 3653-80.

Boateng, J. S., Matthews, K.H., Stevens, H.N.E., Eccleston, G.M. (2008). Wound healing dressings and drug delivery systems: a review. *J. Pharm. Sci.* 97, 2892–923.

Bombardelli, E., Crippa, E., Pifferi G. (1986). Sericoside, a new glicoside in functional cosmetics" - Preprints of the 14th I.F.S.C.C. Congress, Barcellona, Vol II.

Bombardelli, E., Cristoni, A., Morazzoni, P. (1994). Phytosome®s in functional cosmetics. *Fitoterapia,* 65, 387–401.

Bombardelli, E., Curri, S. B., Della Loggia, R., Del Negro, P., Tubaro, A., Gariboldi, P. (1989). Complex between phospholipids and vegetal derivatives of biological interest. *Fitoterapia,* 60, 1–9 [Suppl. to issue N.1].

Bombardelli, E., Curri, S. B., Gariboldi P. (1989). Cosmetic utilization of complexed of Panax Ginseng saponins with phospholipids in the Phytosome® Form. *Fitoterapia,* 60 (Suppl. N.1), 55.

Busby, A., La Grange, L., Edwards, J., King, J. (2002). The use of a silymarin/phospholipid compound as a fetoprotectant from ethanol-induced behavioral deficits. *J. Herb. Pharmacother.,* 2, 39–47.

Buzzelli, G., Moscarella, S., Giusti, A., Duchini, A., Marena, C., Lampertico, M. (1993). A pilot study on the liver protective effect of silybinphosphatidylcholine complex (IdB 1016) in chronic active hepatitis. *Int. J. Clin. Pharmacol. Ther. Toxicol.,* 31, 456–60.

Cacciapuoti, F., Scognamiglio, A., Palumbo, R., Forte, R., Cacciapuoti, F. (2013). Silymarin in non alcoholic fatty liver disease. *World J. Hepatol.,* 5, 109–13.

Chew, F.S. (1988). Biological effects of glucosinolates. In: Cutler, H.G. (Ed.), Biologically Active Natural Products: Potential Use in Agriculture, (pp. 155–181). Washington, DC, American Chemical Society.

Canini, F., Bartolucci, L., Cristallini, E., Gradoli, C., Rossi, A., Ribacchi, R., Valori, C. (1985). L'impiego della silimarina nel trattamento della steatosi epatica alcolica. *Clin. Ter.,* 114, 307–14.

Carroll, R. E., Benya, R. V., Turgeon, D. K., Vareed, S., Neuman, M., Rodriguez, L., Kakarala, M., Carpenter, P. M., McLaren, C., Meyskens, F. L. Jr, Brenner, D. E. (2011). Phase IIa clinical trial of curcumin for the prevention of colorectal neoplasia. *Cancer Prev. Res.* (Phila)., 4, 354–64.

Changediya, V., Khadke, M., Devdhe, S. (2011). Phytosomes: new approach for delivering herbal drug with improved bioavailability. *Res. J. Pharm. Biol. Chem. Sci.,* 2, 57–68.

Cohn, J. S., Wat, E., Kamili, A., Tandy, S. (2008). Dietary phospholipids, hepatic lipid metabolism and cardiovascular disease. *Curr. Opin. Lipidol.,* 19, 257–62.

Conti, M., Malandrino, S., Magistretti, M. J. (1992). Protective activity of silipide on liver damage in rodents. *Jpn. J. Pharmacol.,* 60, 315–21.

Cristoni, A. (2005). Cosmetic applications of natural pentacyclic triterpenes. *Nutracos,* March-April.

Cuomo, J., Appendino, G., Dern, A. S., Schneider, E., McKinnon, T. P., Brown, M. J., Togni, S., Dixon, B. M. (2011). *J. Nat. Prod.,* 74, 664–9.

Dashwood, R. H. (2007). Frontiers in Polyphenols and Cancer Prevention. *J. Nutr.,* 137, 267S–9S.

De Feudis, F. V. (1991). Ginkgo biloba extract (EGb 761): *Pharmacological Activities and Clinical Applications,* Amsterdam, Elsevier.

Disilvestro, R.A. (2001). Flavonoids as Antioxidants. In: Wildman, R.E.C., (Ed.) *Handbook of Nutraceuticals and Functional Foods,* (pp. 127–142). Boca Raton, FL, CRC Press.

Djekic, L., Krajisnik, D., Micic, Z. (2015). Polyphenolics-phospholipid complexes as natural cosmetic ingredients: Properties and application. *Tenside Surfact. Det.,* 52, 186-92.

Djekic, L., Krajišnik, D., Mićic, Z., Čalija, B. (2016). Formulation and physicochemical characterization of hydrogels with 18β-glycyrrhetinic acid/phospholipid complex phytosomes. *J. Drug Deliv. Sci.,* 35, 81–90.

Dorai, T., Aggarwal, B. B. (2004). Role of chemopreventive agents in cancer therapy. *Cancer Lett.,* 215, 129–40.

Du, Y., Guo, H., Lou, H. (2007). Grape seed polyphenols protect cardiac cells from apoptosis via induction of endogenous antioxidant enzymes. *J. Agric. Food Chem.,* 55, 1695–1701.

El-Lakkany, N. M., Hammam, O. A., El-Maadawy, W. H., Badawy, A. A., Ain-Shoka, A. A., Ebeid, F. A. (2012). Anti-inflammatory/anti-fibrotic effects of the hepatoprotective silymarin and the schistosomicide praziquantel against Schistosoma mansoni-induced liver fibrosis. *Parasit Vectors.* 5:9 (http://www.parasitesandvectors.com/content/5/1/9).

Féher, J., Lengyel, G. (2012). Silymarin in the prevention and treatment of liver diseases and primary liver cancer. *Curr. Pharm. Biotechnol.*, 13, 210–7.

Ferenci, P. (2013). Silymarin in the treatment of liver diseases: What is the clinical evidence? *CLD,* 7, 8–10.

Ferenci, P., Dragosics, B., Dittrich, H., Frank, H., Benda, L., Lochs, H., Meryn, S., Base, W., Schneider, B. (1989). Randomized controlled trial of silymarin treatment in patients with cirrhosis of the liver. *J. Hepatol.,* 9, 105–13.

Ferrara, T., De Vincentiis, G., Di Pierro, F. (2015). Functional study on Boswellia phytosome as complementary intervention in asthmatic patients. *Eur. Rev. Med. Pharmacol. Sci.,* 19, 3757-62.

Fried, M. W., Navarro, V. J., Afdhal, N., Belle, S. H., Wahed, A.S., Hawke, R. L., Doo, E., Meyers, C. M., Reddy, K. R. (2012). Silymarin in NASH and C Hepatitis (SyNCH) Study Group. Effect of silymarin (milk thistle) on liver disease in patients with chronic hepatitis C unsuccessfully treated with interferon therapy: a randomized controlled trial. *JAMA,* 308, 274–82.

Gallo, D., Giacomelli, S., Ferlini, C., Raspaglio, G., Apollonio, P., Prislei, S., Riva, A., Morazzoni, P., Bombardelli, E., Scambia, G. (2003). Antitumour activity of the silybin–phosphatidylcholine complex, IdB 1016, against human ovarian cancer. *Eur. J. Cancer*, 39, 2403–10.

Giacomelli, S., Gallo, D., Apollonio, P., Ferlini, C., Distefano, M., Morazzoni, P., Riva, A., Bombardelli, E., Mancuso, S., Scambia, G. (2002). Silybin and its bioavailable phospholipid complex (IdB 1016) potentiate in vitro and in vivo the activity of cisplatin. *Life Sci.,* 70, 1447–59.

Giori, A., Franceschi, F., Patent Application WO 2007/101551.

Godugu, C., Patel, A. R., Doddapaneni, R., Somagoni, J., Singh, M. (2014). Approaches to improve the oral bioavailability and effects of novel anticancer drugs berberine and betulinic acid. *PLoS ONE,* 9, e89919. doi:10.1371/journal.pone.0089919. *eCollection* 2014.

Graham, H. N. (1992). Green tea composition, consumption, and polyphenol chemistry. *Preventive Medicine,* 21, 334–50.

Halliwell, B. (2008). Are polyphenols antioxidants or pro-oxidants? What do we learn from cell culture and in vivo studies? *Arch. Biochem. Biophys.*, 476, 107–12.

Hsu, C. H., Cheng, A. L. (2007). Clinical studies with curcumin. *Adv. Exp. Med. Biol.*, 595, 471–80.

Hüsch, J., Bohnet, J., Fricker, G., Skarke, C., Artaria, C., Appendino, G., Schubert-Zsilavecz, M., Abdel-Tawab, M. (2013). Enhanced absorption of boswellic acids by a lecithin delivery form (Phytosome®) of Boswellia extract. *Fitoterapia,* 84, 89–98.

Kapoor Silki, D., Malviya, S., Talwar, V., Prakash, K. O. (2012). Potential and promises of phospholipid structured novel formulations for hepatoprotection. *Int. J. Drug Dev. Res.,* 4, 51–8.

Katiyar, S. K. (2005). Silymarin and skin cancer prevention: anti-inflammatory, antioxidant and immunomodulatory effects. *Int. J. Oncol.,* 26, 169–76.

Kennedy, D. O., Haskell, C. F., Mauri, P.L., Scholey, A. B. (2007). Acute cognitive effects of standardised Ginkgo biloba extract complexed with phosphatidylserine. *Hum. Psychopharmacol. Clin. Exp.,* 22, 199–210.

Khan, J., Alexander, A., *Ajazuddin,* Saraf, S., Saraf S (2013). Recent advances and future prospects of phyto-phospholipid complexation technique for improving pharmacokinetic profile of plant actives. *J. Control. Release,* 168, 50–60.

Kidd, P. M. (2009). Bioavailability and Activity of Phytosome Complexes from Botanical Polyphenols: The Silymarin, Curcumin, Green Tea, and Grape Seed Extracts. *Altern. Med. Rev.,* 14, 226–46.

Kidd, P. (2002). Phospholipids: versatile nutraceuticals for functional foods. *Functional ingredients.* (December), 1–11.

Kitajima, S., Yamaguchi, K. (2009). Silybin from Silybum marianum seeds inhibits confluent-induced keratynocytes differentiation as effectively as retinoic acid without inducing inflammatory cytokine. *J. Clin. Nutr.,* 45, 178–84.

La Grange, L., Wang, M., Watkins, R., Ortiz, D., Sanchez, M. E., Konst, J., Lee, C., Reyes, E. (1999). Protective effects of the flavonoid mixture, silymarin, on fetal rat brain and liver. *J. Ethnopharmacol.,* 65, 53–61.

Lazarus, G.S., Cooper, D.M., Knighton, D.R., Margolis, D. J., Pecoraro, R. E., Rodeheaver, G., Robson, M. C. (1994). Definitions and guidelines for assessment of wounds and evaluation of healing. *Arch. Dermatol.,* 130, 489–93.

Lee, H., Lee, C., Kim, J., Pyo, S. (2014). The inhibitory effect of sinigrin on the production of inflammatory mediators induced by lipopolysaccharide in RAW 264.7 macrophages. *FASEB J.* 28, no. 1 Supplement, 1056.5.

Li, J., Wang, X., Zhang, T., Wang, C., Huang, Z., Luo, X., Deng, Y. (2015). A review on phospholipids and their main applications in drug delivery systems. *AJPS,* 10, 81–8.

List, G. R. (2015). 1- Soybean Lecithin: Food, Industrial Uses, and Other Applications. *Polar Lipids,* 1–33.

Loguercio, C., Federico, A., Trappoliere, M., Tuccillo, C., De Sio, I., Di Leva, A., Niosi, M., D'Auria, M. V., Capasso, R., Del Vecchio Blanco, C., Real Sud Group. (2007). The effect of a silybin–vitamin E–phospholipid complex on nonalcoholic fatty liver disease: a pilot study. *Dig. Dis. Sci.,* 52, 2387–95.

Manach C., Williamson G., Morand C., Scalbert A., Remesy C. H. (2005). Bioavailability and bioefficacy of polyphenols in humans. *Am. J. Clin. Nutr.,* 81, 230S-42S.

Manach, C., Donovan, J. L. (2004). Pharmacokinetics and metabolism of dietary flavonoids in humans. *Free Radic. Res.,* 38, 771–85.

Mahadevan, S., Park, Y. (2008). Multifaceted therapeutic benefits of Ginkgo biloba L.: chemistry, efficacy, safety, and uses. *J. Food Sci.,* 73, R14–9.

Maiti, K., Mukherjee, K., Gantait, A., Ahamed, H. N., Saha, B. P., Mukherjee, P. K. (2005). Enhanced therapeutic benefit of quercetin–phospholipid complex in carbon tetrachloride-induced acute liver injury in rats: a comparative study. *Iran J. Pharmacol. Ther.,* 4, 84–90.

Mazumder[a], A., Dwivedi, A., Plessis, J.D. (2016). Sinigrin and its therapeutic benefits. *Molecules,* 21, 416.

Mazumder[b], A., Dwivedi, A, Fox, L. T., Brümmer, A., du Preez, J. L., Gerber, M., du Plessis, J. (2016). In vitro skin permeation of sinigrin from its phytosome complex. *J. Pharm. Pharmacol.,* Oct 2. doi: 10.1111/jphp. 12594. [Epub ahead of print]

Morazzoni, P., Montalbetti, A., Malandrino, S., Pifferi, G. (1993). Comparative pharmacokinetics of silipide and silymarin in rats. *Eur. J. Drug Metab. Pharmacokinet.,* 18, 289–97.

Morazzoni, P., Petrini, O., Scholey, A., Kennedy, D. (Indena S.p.A., Boehringer Ingelheim Int.). Use of a Ginkgo complexes for the enhancement of cognitive functions and the alleviation of mental fatigue. Patent WO2005074956A1, 2005.

Muir, A. H., Robb, R., McLaren, M., Daly, F., Belch, J. J. F. (2002). The use of Ginkgo biloba in Raynaud's disease: a doubleblind placebo-controlled trial. *Vasc. Med.,* 7, 265–7.

Marena, C., Lampertico, M. (1991). Preliminary clinicaldevelopment of silipide: a new complex of silybin in toxic liver disorders. *Planta Med.,* 57, A124–5.

Myiata, S., et al. (2005). Effects of silybin (milk thistle extract) on the skin antiageing Action. IFSCC Congress.

Naik, S. R., Panda, V. S. (2008). Hepatoprotective effect of Ginkgoselect Phytosome® in rifampicin induced liver injurym in rats: Evidence of antioxidant activity. *Fitoterapia,* 79(6), 439–445.

Natella, F., Belelli, F., Gentili, V., Ursini, F., Scaccini, C. (2002). Grape seed proanthocyanidins prevent plasma postprandial oxidative stress in humans. *J. Agric. Food Chem.* 50, 7720–5.

Nuttal, S. L., Kendall, M. J., Bombardelli, E., Morazzoni, P. (1998). An evaluation of the antioxidant activity of a standardized grape seed extract, Leucoselect. *J. Clin. Pharm. Ther.* 23, 385–9.

Oliyarnyk, O., Malinska, H., Trnovska, J., Skop, V., Vecer, R. (2014). Effect of micronized extract and phytosomes containing extract from silybum marianum on metabolic disordes in nonobese model of insulin resistance. Abstracts / *Atherosclerosis,* 235, e191.

Panda, V. S., Naik, S. R. (2008). Cardioprotective activity of Ginkgo biloba Phytosomes in isoproterenol-induced myocardial necrosis in rats: A biochemical and histoarchitectural evaluation. *Exp. Toxicol. Pathol.,* 60, 397–404.

Percival, J. (2002). Classification of wounds and their management. *Surgery,* 20, 114–7.

Pradhan, S. C., Girish, C. (2006). Hepatoprotective herbal drug, silymarin from experimental pharmacology to clinical medicine. *Indian J. Med. Res.,* 124, 491–504.

Quiñones, M., Miguelhttp://www.sciencedirect.com/science/article/pii/ S1043 661812002058 - cor0005mailto:marta.miguel@csic.es, M., Aleixandre, A. (2013). Beneficial effects of polyphenols on cardiovascular disease. *Pharm. Res.,* 68, 125–31.

Rechner, A. R., Kuhnle, G., Bremner, P., Hubbard, G. P., Moore, K. P., Rice-Evans, C. A. (2002). The metabolic fate of dietary polyphenols in humans. *Free Radic. Biol. Med.,* 33, 220–35.

Rungapamestry, V., Rabot, S., Fuller, Z., Ratcliffe, B., Duncan, A.J. (2008). Influence of cooking duration of cabbage and presence of colonic microbiota on the excretion of N-acetylcysteine conjugates of allyl isothiocyanate and bioactivity of phase 2 enzymes in F344 rats. *Br. J. Nutr.,* 99, 773–81.

Saller, R., Meier, R., Brignoli, R. (2001). The use of silymarin in the treatment of liver diseases. *Drugs,* 61, 2035–63.

Salmi, H. A., Sarna, S. (1982). Effect of silymarin on chemical, functional and morphological alterations of the liver: a double-blind controlled study. *Scand. J. Gastroenterol.,* 17, 517–21.

Scholfield, C. R. (1981). Composition of soybean lecithin. *J. Am. Oil Chem. Soc.,* 58, 889–92.

Semalty, A., Semalty, M., Rawat, M. S., Franceschi, F. (2010). Supramolecular phospholipids–polyphenolics interactions: The Phytosome® strategy to improve the bioavailability of phytochemicals. *Fitoterapia,* 81, 306–14.

Shi, J., Yu, J., Pohorly, J. E., Kakuda, Y. (2003). Polyphenolics in grape seeds-biochemistry and functionality. *J. Med. Food* 6, 291–9.

Shofran, B.G., Purrington, S., Breidt, F., Fleming, H. (1998). Antimicrobial properties of sinigrin and its hydrolysis products. *J. Food Sci.,* 63, 621–4.

Siddiqui, M. Z. (2011). Boswellia Serrata, A Potential Antiinflammatory Agent: An Overview. *Indian J. Pharm. Sci.,* 73, 255–61.

Siemoneit, U., Koeberle, A., Rossi, A., Dehm, F., Verhoff, M., Reckel, S., Maier, T. J., Jauch, J., Northoff, H., Bernhard, F., Doetsch, V., Sautebin, L., Werz, O. (2011). Inhibition of microsomal prostaglandin E2 synthase-1 as a molecular basis for the anti-inflammatory action of boswellic acids from frankincense. *Br. J. Pharmacol.,* 162, 147–62.

Singh, A., Anand Saharan, V., Singh, M., Bhandari, A. (2011). Phytosome: Drug Delivery System for Polyphenolic Phytoconstituents. *IJPS,* 7, 209–19.

Singh, R.P. Agarwal, R. (2009). Cosmeceuticals and silibinin. *Clin. Dermatol.,* 27, 479–84.

Singhal, M., Yashwant, Nayak, A., Singh, V., Singh parihar, A. (2009). Curcumin: A Chemopreventive Agent in Pre-malignant Lesions. *IJTPR,* 1, 27–32.

Siviero, A., Gallo, E., Maggini, V., Gori, L., Mugelli, A., Firenzuoli, F., Vannacci, A. (2015). Curcumin, a golden spice with a low bioavailability. *J. Herb. Med.,* 5, 57–70.

Skarke, C., Kuczka, K., Tausch, L., Werz, O., Rossmanith, T., Barrett, J. S., Harder, S., Holtmeier, W., Schwarz, J. A. (2012). Increased bioavailability of 11-keto-β-boswellic acid following single oral dose frankincense extract administration after a standardized meal in healthy volunteers: modeling and simulation considerations for evaluating drug exposures. *J. Clin. Pharmacol.*, 52, 1592–600.

Tangney, C. Rasmussen, H. E. (2013). Polyphenols, Inflammation, and Cardiovascular Disease. *Curr. Atheroscler. Rep.*, 15, 324. doi: 10.1007/s11883-013-0324-x.

Tausch, L., Henkel, A., Siemoneit, U., Poeckel, D., Kather, N., Franke, L., Hofmann, B., Schneider, G., Angioni, C., Geisslinger, G., Skarke, C., Holtmeier, W., Beckhaus, T., Karas, M., Jauch, J., Werz, O. (2009). Identification of human cathepsin G as a functional target of boswellic acids from the anti-inflammatory remedy frankincense. *J. Immunol.*, 183, 3433–42.

Tedesco, D., Steidler, S., Galletti, S., Tameni, M., Sonzogni, O., Ravarotto, L. (2004). Efficacy of silymarin–phospholipid complex in reducing the toxicity of aflatoxin B1 in broiler chicks. *Poult. Sci.*, 83, 1839–43.

Tsao, R. (2010). Chemistry and Biochemistry of Dietary Polyphenols. *Nutrients* 2, 1231–46.

Valenzuela, A., Aspillaga, M., Guerra, R. (1989). Selectivity of silamarin on the increase of glutathione containing in different tissues of Rat. *Plant Med.*, 55, 420–2.

Vigna, G. B., Costantini, F., Aldini, G., Carini, M., Catapano, A., Schena, F., Tangerini, A., Zanca, R., Bombardelli, E., Morazzoni, P., Mezzetti, A., Fellin, R., Maffei Facino, R. (2003). Effect of a standardized grape seed extract on low-density lipoprotein susceptibility to oxidation in heavy smokers. *Metabolism*, 52, 1250–7.

WHO. Guidelines on Developing Consumer Information on Proper Use of Traditional, Complementary and Alternative Medicine. *World Health Organization*, 2004.

Whittam, A. J., Maan, Z. N., Duscher, D., Wong, V.W., Barrera, J. A., Januszyk, M., Gurtner, G. C. (2016). Challenges and Opportunities in Drug Delivery for Wound Healing. *Adv. Wound Care*, 5, 79–88.

Williamson, G., Barron, D., Shimoi, K., Terao, J. (2005). In vitro biological properties of flavonoid conjugates found in vivo. *Free Radic. Res.*, 39, 457–69.

Xia, E.-Q., Deng, G.-F., Guo, Y.-J., Li, H.-B. (2010). Biological Activities of Polyphenols from Grapes. *Int. J. Mol. Sci.*, 11, 622–46.

Yu, F., Li, Y., Chen, Q., He, Y., Wang, H., Yang, L., Guo, S., Meng, Z., Cui, Z., Xue, M., Chen, X. D. (2016). Monodisperse microparticles loaded with the self-assembled berberine-phospholipid complex-based phytosomes for improving oral bioavailability and enhancing hypoglycemic efficiency. *Eur. J. Pharm. Biopharm.*, 103, 136–148.

Zhang, J., Tang, Q., Xu, X., Li, N. (2013). Development and evaluation of a novel phytosome-loaded chitosan microsphere system for curcumin delivery. *Int. J. Pharm.*, 448, 168–74.

Zhou, B., Wu, L.-M., Yang, L., Liu, Z.-L. (2005). Evidence for alpha-tocopherol regeneration reaction of green tea polyphenols in SDS micelles. *Free Radic. Biol. Med.*, 38, 78–84.

BIOGRAPHICAL SKETCH

Ljiljana Djekic

University of Belgrade, Faculty of Pharmacy,
Department of Pharmaceutical technology and Cosmetology,
Vojvode Stepe 450, 11221 Belgrade, Serbia

Education:

- 2013: Specialization in Pharmaceutical technology
- 2011: PhD degree in pharmaceutical sciences
- 2004: MS degree in pharmaceutical sciences
- 1998: BS degree
- 1992-1998: Faculty of Pharmacy, University of Belgrade

Research and Professional Experience:

- 2012-present: Assistant Professor, Department of Pharmaceutical Technology and Cosmetology, Faculty of Pharmacy, University of Belgrade
- 2004-2012: Assistant, Department of Pharmaceutical Technology and Cosmetology, Faculty of Pharmacy, University of Belgrade
- 1998-2004: Teaching assistant, Department of Pharmaceutical Technology and Cosmetology, Faculty of Pharmacy, University of Belgrade

Professional Appointments:

- 2012-present Assistant Professor, Department of Pharmaceutical Technology and Cosmetology, Faculty of Pharmacy, University of Belgrade
- 2004-2012 Assistant, Department of Pharmaceutical Technology and Cosmetology, Faculty of Pharmacy, University of Belgrade
- 1998-2004 Teaching assistant, Department of Pharmaceutical Technology and Cosmetology, Faculty of Pharmacy, University of Belgrade

Honors:

- Recognized Reviewer Status awarded in April 2015 by Elsevier, International Journal of Pharmaceutics.
- Outstanding Reviewer Status awarded in July 2015 by Elsevier, International Journal of Pharmaceutics.
- Reviewer of the papers submitted for presentation on the International Conference on Material Technology and Environmental Engineering (MTEE 2015), August 14-15, 2015, Shanghai, China.
- Invited Peer Reviewer of the The National Foundation for Science and Technology Development, Ha Noi, Vietnam, for the project proposals under the basic research funding program in natural sciences and enginering 2015.
- Reviewer of a numerous scientific papers for the journals: International Journal of Pharmaceutics; Journal of Biomaterials Science: Polymer Edition; Colloids and Surfaces B: 2/4 Biointerfaces, Drug Development and Industrial Pharmacy; Journal of Nanobiotechnology; Current Drug Delivery; Journal of Drug Delivery Science and Technology; Journal of the Serbian Chemical Society; Chiang Mai Journal of Science; The Journal of Pharmaceutics & Drug Delivery Research; Journal of Pharmaceutics & Drug Development (JPDD); Arhiv za farmaciju.
- Awarded with International Poster Award, 62nd International Congess of International Pharmaceutical Federation, Nice, France, 2002.
- The active membership of Pharmaceutical Association of Serbia (2010 - present)

Publications last three years:

[1]   Djekic L. Liposomes: Properties and Therapeutic Applications. In: Novel Approaches for Drug Delivery. Raj K. Keservani, Anil K. Sharma, Rajesh Kumar Kesharwani (Eds). IGI Global, Hershey 2017, pp. 27 – 51 (DOI: 10.4018/978-1-5225-0751-2.ch002) (In press).

[2]   Djekic L, Primorac M. Microemulsions and nanoemulsions as carriers for delivery of NSAIDs. In: Microsized and Nanosized Carriers for Nonsteriodal Anti-Inflammatory Drugs: Potential benefits and challenges. Čalija Bojan (Ed). Elsevier, Philadelphia, 2016 (ISBN 9780128040171) (In production).

[3]   Djekic L, Primorac M. Percutaneous Penetration Enhancement Potential of Microemulsion-Based Organogels. In: Percutaneous Penetration Enhancers. Chemical Methods in Penetration Enhancement. Nanocarriers. Nina Dragicevic, Howard I. Maibach (Eds), Springer-Verlag, Berlin, Heidelberg, 2016, pp. 263 - 282. (DOI 10.1007/978-3-662-47862-2_17) (ISBN 978-3-662-47861-5).

[4]   Djekic L, Primorac M. Biomedical Application of Fullerenes. Fullerenes: Chemistry, Natural Sources and Technological Applications (Shannon B. Ellis ed.), Nova Science Publishers, Hauppauge, 2014, pp. 239-262. (ISBN: 978-1-63321-386-9).

[5]   Djekic L, Martinovic M, Stepanović-Petrović R, Micov A, Tomić M, Primorac M. Formulation of hydrogel-thickened nonionic microemulsions with enhanced percutaneous delivery of ibuprofen assessed in vivo in rats. Eur J Pharm Sci, In Press, Corrected Proof, Available online 6 May 2016 (doi:10.1016/j.ejps.2016.05.005).

[6]   Djekic L, Krajišnik D, Micic Z, Čalija B. Formulation and physicochemical characterization of hydrogels with 18ß-glycyrrhetinic acid/phospholipid complex phytosomes. Journal of Drug Delivery Science and Technology, 35 (2016) 81 – 90, Available online 16 Jun 2016 (DOI: 10.1016/j.jddst.2016.06.008).

[7]   Janković J, Djekic L, Dobričić V, Primorac M. Evaluation of critical formulation parameters in design and differentiation of self-microemulsifying drug delivery systems (SMEDDSs) for oral delivery of aciclovir. Int J Pharm. 2016 Jan 30;497(1-2):301-11. doi: 10.1016/j.ijpharm.2015.11.011. Epub 2015 Dec 1.

[8] Djekic L, Martinovic M, Stepanović-Petrović R, Tomić M, Micov A, Primorac M. Design of Block Copolymer Costabilized Nonionic Microemulsions and Their In Vitro and In Vivo Assessment as Carriers for Sustained Regional Delivery of Ibuprofen via Topical Administration. J Pharm Sci. 2015 Aug;104(8):2501-12.

[9] Djekic L, Krajisnik D, Martinovic M, Djordjevic D, Primorac M. Characterization of gelation process and drug release profile of thermosensitive liquid lecithin/poloxamer 407 based gels as carriers for percutaneous delivery of ibuprofen. Int J Pharm. 2015 Jul 25;490(1-2):180-9.

[10] Krstić M, Ražić S, Djekić Lj, Dobričić V, Momčilović M, Vasiljević D, Ibrić S. Application of a Mixture Experimental Design in the Optimization of the Formulation of Solid Self-Emulsifying Drug Delivery Systems with Carbamazepine. Latin American Journal of Pharmacy. 2015; volumen 34, issue 5.

[11] Ljiljana Djekic, Danina Krajisnik, Zorica Micic. Polyphenolics-Phospholipid Complexes as Natural Cosmetic Ingredients: Properties and Application. Tenside Surf. Det. 52 (2015) 186-192.

[12] Ljiljana M. Djekić, Marija M. Primorac. Formulacija i karakterizacija samo-mikroemulgujućih nosača lekovitih supstanci na bazi biokompatibilnih nejonskih surfaktanata. Hem. Ind. 68 (5) 565–573 (2014).

In: Polyphenolics
Editor: Patricia Clark

ISBN: 978-1-53610-709-8
© 2017 Nova Science Publishers, Inc.

Chapter 3

HEALTH-PROMOTING EFFECTS OF POLYPHENOLS FROM *MORINGA OLEIFERA* LAM AGAINST NON-COMMUNICABLE DISEASES

Zaina B. Ruhomally and *Vidushi S. Neergheen-Bhujun*[*], *PhD*

Department of Health Sciences and ANDI Centre of Excellence for Biomedical and Biomaterials Research Faculty of Science, University of Mauritius, Reduit, Republic of Mauritius

ABSTRACT

Moringa oleifera Lam, a very resilient plant, currently cultivated in all the tropical and sub-tropical regions of the world has a rich history as traditional medicine and food. *M. oleifera* is being used to improve the nutritional value of staple foods in many parts of the world including Africa. It is considered as an important food forticant in view of the appreciable amounts of proteins, carbohydrates, starch as well as specific micronutrients like calcium, iron, magnesium, manganese, copper and potassium particularly in the leaves. Likewise, an impressive range of intrinsic bioactive phytonutrients with known antioxidant promises such as ascorbic acid, phenolics, flavonoids, tocopherols and carotenoids have

---

[*] Corresponding author: v.neergheen@uom.ac.mu.

also been recognized in this miracle tree. In addition, to its high nutritive values and powerful water purifying abilities, it is being considered for the prevention or treatment of a number of non-infectious ailments because of its pluripharmacological properties ranging from antioxidants, anti-inflammatory, anti-hypertensive to anticancer. These properties can be ascribed to the host of polyphenolic compounds present in these plants. Reported studies conducted on experimental animals and on human cell lines, though inadequate in number, seem concordant in their support for these properties. Therefore, the nutritional, prophylactic and therapeutic potentials of this multipurpose plant are being commended and this has allowed to envisage their potential applications as fresh produce or dried as a functional food and nutraceutical health promoter. This chapter will therefore, explore current scientific data on the therapeutic potential of the phenolic phytonutrients from *M. oleifera* plants in relation to the management of non-communicable diseases.

1. INTRODUCTION

Despite the major advancements in the standards of health care and medical technology in developed as well as in developing countries, non-communicable diseases (NCDs) like cardiovascular disease, hypertension, diabetes, obesity, Alzheimer's disease, cancer amongst others, are globally considered as major health threats that can translate into extensive socio-economic problems (Ng et al. 2014). In fact, according to the World Health Organization (WHO) 2014 report, NCDs kill 38 million of people each year and approximately 28 million of these deaths occur in low and middle income countries. Cardiovascular diseases account for most NCD deaths annually (17.5 million) followed by cancers (8.2 million), respiratory diseases (4 million), and diabetes (1.5 million) (WHO, 2014). Also, adulthood obesity prevalence predictions (2010–2030) forecast that in 2020, 44% of men and 31% of women will be obese (WHO, 2013). At this precipitous rise in the incidence of these diseases, NCDs appear as a major global challenge faced by limited new treatment coming on the market and drug resistance in other cases. The diseases are marked by disrupted cellular and molecular pathways, which are highly relevant as targets for phytotherapeutics.

Nature harbors a wide variety of flora and fauna, which provides an excellent source of chemically diverse compounds with great therapeutic potential (Batnagar and Kim, 2010). Natural products (NPs) and their derivatives have historically been used as therapeutic agents (Koehn and Carter, 2005). During the last few years, extensive research was carried out to search for novel compounds from natural products, which has become candidates for drug discovery. For instance, more than 60% of the drugs on the market are derived from natural sources (Molinari, 2009). These natural resources in particular traditionally used plants provide scope for discovery of novel pharmaceutical agents. The promotion of healthy diets, nutrition and lifestyles to thwart the global burden of these diseases is also constantly being advocated, and a promising strategy is the copious consumption of nutraceuticals, functional foods and value added food products, due to their easy availability and diseases averting characteristics. Functional foods are those that are thought to have physiological benefits and/or reduce the risk of chronic disease beyond their basic nutritional functions and are rich repositories of health promoting bioactive compounds. Research reveal that a large proportion of these compounds often act as antioxidants, and have a role *in–vivo* in modulating disease development by inhibiting ROS mediated reactions, which have been associated with the initiation and progression of a number of pathological processes (Bajpai et al., 2005).

This book chapter therefore, focuses on the role of phytochemicals rich *M. oleifera* on multiple targets in the pathophysiological pathways involved in non-communicable diseases. The chapter also highlights the protective effects of the bioactive constituents from *M. oleifera* with the scope of developing novel therapeutic targets to manage and eventually mitigate non-communicable diseases.

1.1. *Moringa oleifera*

Moringa oleifera (Lam) from the Moringaceae family has long been used for culinary purposes and as traditional medicine. This resilient plant grows well at altitudes from 0 to 1,800 m and in areas with rainfall between 500 and 1,500 mm per year, making it suitable for both semi-arid and arid ecosystem, which covers 37.0% of the earth's geographical area, and even larger swaths of the developing world (Yang et al., 2015).

1.2. Traditional Uses

M. oleifera leaves, pods and flowers have been reported both for their use as food and medicine. Succulent leaves are used in soups, sauces, porridges or salads (Lockett et al., 2000). In many cultures it is also used as curry gravy in noodles, rice or wheat (Abilgos et al., 1999). Fresh leaves can also be eaten cooked and are often, shade dried, ground to a powder, and then stored for later use as a food flavoring or additive without any major loss in their nutritional value. Farmers have added the leaves to animal feed to maintain a healthy livestock (Sarwatt et al., 2002; Fahey, 2005; Sáncheza et al., 2006). Newer applications include the use of Moringa powder as a fish food in aqua cultural systems (Dongmeza et al., 2006) and the leaves as a protein supplement for cows since the feeding value of Moringa has been reported to be similar to that of soybeans and rapeseed meal (Soliva et al., 2005). In Mauritius, *M. oleifera* has a long-standing culinary tradition. It can be obtained from neighbours or friends who have the tree in their backyard. Usually the fresh leaves are used in fish soups and "bouillon" or can be sautéed. The pods are mainly used in many traditional dishes such as drumstick fish curry and can also be used in soups (Nazim et al., 2015).

In underdeveloped countries pregnant women and lactating mothers use the powdered leaves to enhance their child nourishment (Hassan and Ibrahim, 2013). In the Philippines, Moringa is known as 'mother's best friend' because of its utilization to increase woman's milk production (Kumar et al., 2010). *M. oleifera* was also massively grown and promoted by the local media in Uganda in the 1980s as a plant putatively able to cure a number of diseases including symptoms of HIV/AIDS (Kasolo et al., 2010). Moreover, Moringa oil has been used in skin ointments ever since the Egyptian times (Razis et al., 2014).

Consequently, different part of the tree has been used in the treatment of a wide variety of ailments. The tree continues to have an important role particularly as counter-irritant in the indigenous medicine in Asia and West Africa (Arora et al., 2013). The root, bark, gum, leaf, fruit (pods), flowers, seed and seed oil have been used against various diseases in the indigenous medicine of South Asia, including the treatment of inflammation and infectious diseases along with cardiovascular, gastrointestinal, haematological and hepatorenal disorders (Kumar et al., 2010). Furthermore, *M. oleifera* has long been recognized in various traditional medicinal systems in the treatment and management of an array of non-communicable diseases including the treatment of inflammation, cardiovascular disease, anemia, swelling, anxiety, asthma, pain, abnormal blood pressure, and haematological disorders among others (Ndhlala et al., 2014; Khawaja et al., 2010). Moreover, it was reported that fresh juice of *M. oleifera* leaves mix with milk were consumed by people of Mauritius for the management of type 2 diabetes and high cholesterol levels. Likewise, the decoction of the root and stem were also consumed to manage hypertension (Mahomoodally et al., 2016).

1.3. Nutritional Value of *Moringa oleifera*

M. oleifera is an important food commodity which has enormous attention as the 'natural nutrition of the tropics'. The leaves, fruit, flowers, and immature pods of this tree are used as a highly nutritive vegetable in many countries, particularly in India, Pakistan, Philippines, Hawaii, and many parts of Africa (Anwar et al., 2007). It was reported that fresh Moringa leaves contain about 7 times the vitamin C of oranges, 4 times the calcium of milk, 4 times the vitamin A of carrots, 3 times the potassium of bananas, 2 times the protein of yogurt and ¾ times the iron of spinach gram-for-gram (Trees for life, 2005). Likewise, the leaves consist of a fairly unique amount of protein by dry weight and all essential amino acids (approximately 27%) which are unusual in plant sources (Waterman et al., 2014). Apart from being a rich source of B vitamins, the leaves also contain a treasure trove of minerals like calcium (3.65%), zinc (13.03 mg/kg), manganese (86.8 mg/kg), iron (490 mg/kg) and selenium (363 mg/kg) as well as 17 different fatty acids, with α-linolenic acid (44.57%) having the highest value followed by heneicosanoic acid (14.41%). The dried leaves possessed vitamin E (77 mg/100 g) as well as beta-carotene (18.5 mg/100g) (Moyo et al., 2011; Trees for life, 2005). As commonly known, most vegetables lose their nutrients upon cooking. However, it was observed that Moringa leaves whether fresh, cooked, or stored as dried powder for months without refrigeration, did not lose its nutritional value. The leaves which were boiled resulted in three times more bio-available iron than the raw leaves. Similar results were seen in the powdered Moringa leaves (Zaku et al., 2015).

1.4. Phytochemical Constituents from this Miracle Tree

Besides the macronutrient and micronutrient, all parts of *M. oleifera* have been reported to be rich source of phytochemicals. Phytochemicals which are defined as non-nutritive compounds found in plants are of particular interest. These phytochemicals are intermediates or end products of secondary metabolism having protective and therapeutics potential in the management of several diseases (Doughari et al., 2009). They can be classified into four major groups: (i) phenolics and polyphenolics, (ii) terpenoids, (iii) nitrogen-containing alkaloids and (iv) sulphur-containing compounds (Crozier et al., 2006). Reflecting a recent trend of research in natural products, an increasing number of phenolic compounds have been reported with fascinating prophylactic properties. Polyphenols are the most abundant antioxidants in our diet comprising of molecules with hydroxyl groups (D'Archivio et al., 2007). This class of plant metabolites contains more than 8000 known compounds including the main groups: flavonoids, phenolic acids, phenolic alcohols, stilbenes and ligans (Tsao, 2010; D'Archivio et al., 2007). Commonly identified polyphenolic compounds from *M. oleifera* include p-hydroxybenzyl glucosinolates, kaempferol, quercetin, myricetin, isotrifolin, niazirin, moringine and benzoic acid (Figure 1).

Research has revealed the presence of phytochemical constituents like alkaloids, phenolics, flavonoids, glycosides, saponins, terpenoids, tannins, carbohydrates and proteins in various solvent extract of its morphological parts such as the leaves, seeds, flowers, stem and pods (Patel et al., 2014). Accordingly, the antioxidant activities of the leaves of *M. oleifera* has been attributed to the occurrence of a host of flavonoid pigments such as kaempferol, rhamnetin, isoquercitrin, and kaempferitrin (Iqbal and Bhanger, 2006). Flavonol quercetin at concentrations as high as 100 mg/100 g, mainly as quercetin-3-*O*-β-d-glucoside also recognized as isotrifolin or isoquercitrin was identified in dried *M. oleifera* leaves (Atawodi et al., 2010; Lako et al., 2007). In particular, this plant family is also rich in compounds containing the simple sugar, rhamnose, glucosinolates and isothiocyanates (Imdadul et al., 2016).

In addition, glucosinolates are the secondary metabolites that indicate the characteristic of Brassicales plant including *M. oleifera* (Karim et al., 2016). For example, specific components of Moringa preparations have been reported to contain high amount of aromatic glucosinolates namely p-hydroxybenzyl glucosinolates (sinalbin), 2-phenylethyl glucosinolates (gluconasturtiin), benzyl glucosinolates (glucotropaeolin) (Figure 2) (Forster et al., 2015). Seed and leaf of *M. oleifera* have the highest glucosinolates content compared to other parts of plant (Ghosh, 2013). It was revealed that the functional group presented in the compound is commonly responsible for the bioactivities triggered (Karim et al., 2016).

Moreover, isothiocyanates are formed from their glycosylated precursors, glucosinolates via an enzymatic reaction carried out by myrosinase (thioglucoside glucohydrolase), cleaving the thio-linked glucose, leaving the aglycone which repositions rapidly to form the active isothiocyanate. Despite well documented health benefits of isothiocyanates from crucifers, such as anti-inflammatory and anticancer properties, their clinical and dietary use is limited due to their inherent instability. However, Moringa glucosinolates can be converted in situ to four bioactive and relatively stable Moringa isothiocyanates owing to the presence an extra sugar moiety in the aglycone/

isothiocyanate portion of the molecule, which is likely responsible for their high stability (Waterman et al., 2014; Brunelli et al., 2010).

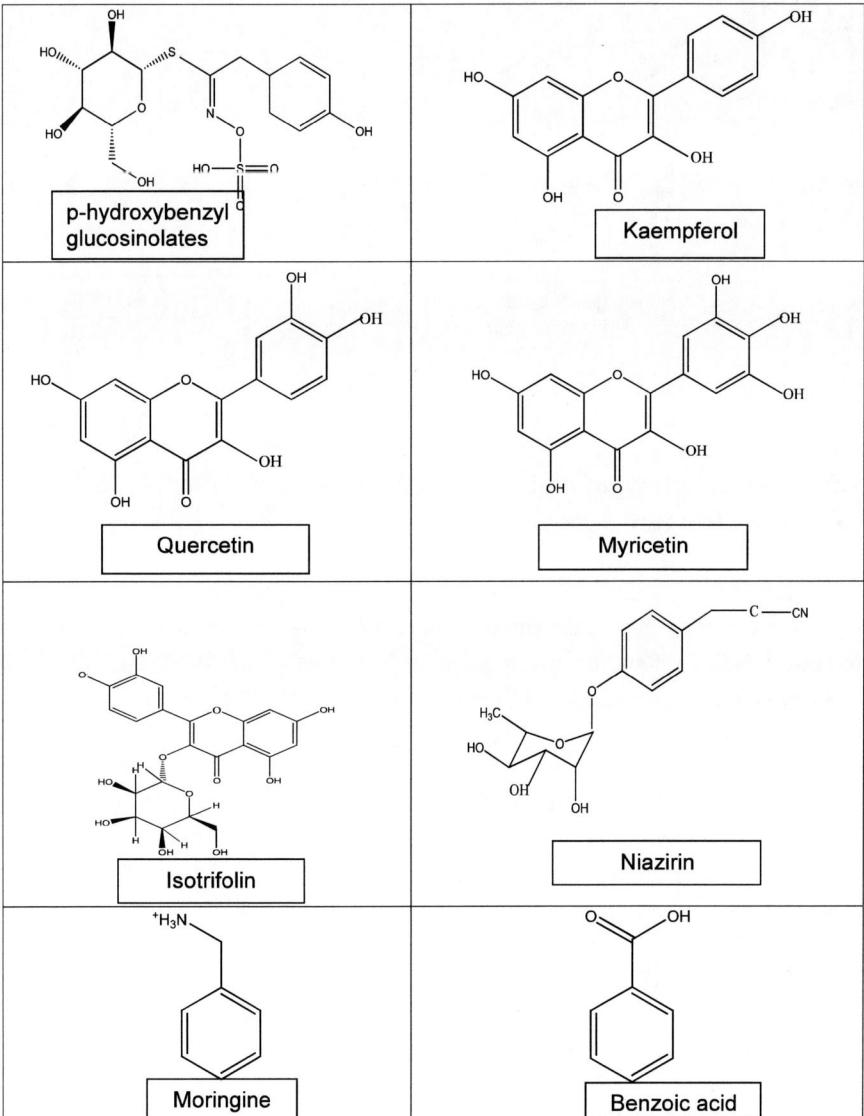

Figure 1. Polyphenolic compounds from *M. oleifera*.

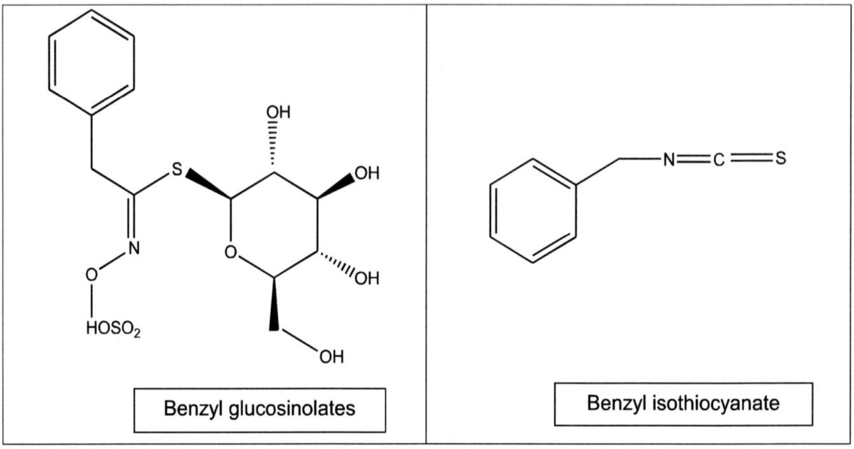

Figure 2. Structure of Glucosinolates and Isothiocyanates from *M. oleifera*.

1.5. Pharmacological Activities of *M. oleifera* Linked to Non-Communicable Diseases

1.5.1. Antioxidant Activity

The viability and functionality of a cell fairly depends on a favorable redox state, for example, on its capacity to thwart excessive oxidation of its macromolecules, comprising DNA, lipids, and proteins (Limon-Pacheco and Gonsebatt, 2009). Highly reactive free radicals and oxygen species are present in biological systems from a wide variety of reactions. Besides their physiological role, these free radicals are ascertained to play a paramount role in the development of a range of oxidative stress-related diseases such as atherosclerosis, ischemic heart disease, cancer, AD, Parkinson's disease and other degenerative diseases (Shahbudin et al., 2011; Valko et al., 2006). Thus, the balance between the production and neutralization of reactive oxygen species (ROS) and reactive nitrogen species (RNS) is vital (Wiernsperger, 2003). This has led to the hypothesis that exogenous dietary antioxidants may have a beneficial role in combating oxidative stress.

Subsequently, phytoantioxidants have garnered enormous attention in modern medicine as well as in traditional system due to their strong antioxidant activity (Kumar and Khanum, 2012). Natural extracts rich in antioxidants are generally recognized to exert protection by inhibiting cellular and molecular damage by free radicals thereby reducing radical-induced tissue injury to finally delay the progress of chronic diseases (Balboa et al., 2013). A recurring justification for the therapeutic actions of *M. oleifera* usage is the relatively strong antioxidant activity of its leaves, flowers, and seeds (Sreelatha and Padma, 2009; Atawodi et al., 2010). It was revealed that the aqueous extract of both mature and tender leaves of *M. oleifera* demonstrated potent scavenging activity on 2, 2-diphenyl-2-picryl hydrazyl (DPPH), superoxide, nitric oxide radical and inhibition of lipid peroxidation (Bholah et al., 2015). Thus, these extracts may prevent oxidative damage to major biomolecules (Sreelatha and Padma, 2009) and reduce the risk of NCDs.

Moreover, a wide range of mechanisms of action of phytochemicals have been proposed. They can either act as antioxidants, or may modulate gene expression and signal transduction pathways upregulating endogenous antioxidant enzymes or molecules (Kris-Etherton et al., 2002; Manson, 2003; Surh, 2003). Phenolics are known to be highly effective antioxidants whereby the protective effects of flavonoids in biological systems are attributed to their ability to scavenge free radicals, chelation of metal ions, trigger antioxidant enzymes, lessen alpha-tocopherol radicals, and impede oxidases (Heim et al., 2002). Their unique chemical structures that are characterized by an aromatic ring consisting of one or more hydroxyl substituents are predictive of their antioxidant effects due to the excellent hydrogen and electron donating capacities. Flavonoids contain a C6-C3-C6 flavon skeleton in which the three-carbon bridge is cyclised with oxygen. The latter antioxidant activity is related to the acid moiety and the number and relative positions of hydroxyl groups on the aromatic ring structure (Bilto et al., 2012).

1.5.2. Anti-Inflammatory Activity

Inflammation underlies a variety of diseases comprising cardiovascular diseases, chronic inflammatory diseases and cancer. The intake of fruit and vegetables is inversely linked with the risk of inflammation and this might correlate intrinsically to the nutrients as well as the non-nutritive bioactive components (Holt et al., 2009). Concomitantly, data from human intervention studies suggested an anti-inflammatory potential of plant-based diet (Watzl, 2008). Intrinsic bioactive compounds mainly carotenoids and flavonoids appear to modulate inflammatory processes. It was reported that phenolics and flavonoids content of plants are responsible for anti-inflammatory activities by inducing free radical scavenging activity and reducing inflammatory cytokines (Ravipati et al., 2012). Moreover, the inhibition of pro-inflammatory enzymes namely nitric oxide synthase (NOS) and cycloxygenase-2 (COX-2) which are mainly responsible for elevating prostaglandin level can be effective targets (Mohanty et al., 2015).

Similarly, it was revealed that aqueous *M. oleifera* root extract at 750 mg/kg reduces carrageenan induced oedema to similar extent as the potent anti-inflammatory drug, indomethacin in rat-paw. Moreover, these results provide further evidence that the roots of *M. oleifera* contain anti-inflammatory principles that may be useful in the treatment of the acute inflammatory conditions (Ndiaye et al., 2002). Likewise, in carrageenan induced paw oedema in rats, both ethanolic and aqueous extract of *M. oleifera* seeds demonstrated dose dependent decrease in paw oedema compared to control group. Thus, these studies are conducive of the anti-inflammatory effects of Moringa plant (Yunus et al., 2013).

This anti-inflammatory activity of *M. oleifera* seeds has been ascribed to innumerable phytochemicals such as alkaloids, flavonoids, sterols, glycosides, tannins and terpenoids (Mahajan et al., 2007; Li and Sinclair, 2002). Sterols like β-sitosterol and flavonoids found in *M. oleifera* seeds are also recognized to target prostaglandin synthesis, which is involved in acute inflammation (Raj et al., 2001). The presence of glycosides like benzyl isothiocyanate in *M. oleifra* seeds have been reported to possess anti-inflammatory activity by declining numerous mediators of inflammation like prostaglandins, nitric oxide, IL-6, IL-1β and TNF-α which are responsible for oedema formation (Lee et al., 2009). Another, probable mechanism could be due to inhibition of free radicals generated like nitric oxide, hydroxyl radicals and superoxide anions which are involved in stimulating inflammatory response (Mohanty et al., 2015). Moreover, aurantiamide acetate, isolated from *M. oleifera* roots and structurally identified as N-benzoylphenylalanyl phenylalinol acetate at 25 μM, inhibited by nearly 90% the secretion TNFα and IL-2 from lipopolysaccharide-stimulated peripheral blood lymphocytes in culture (Sashidhara et al., 2009). Subsequently, this inhibitory activity may contribute to the anti-inflammatory properties of the plant.

1.5.3. Angiotensin I Converting Enzyme (ACE) Inhibition

Novel enzyme inhibitors from plant sources have received great attention due to the copious deleterious effect associated with synthetic drugs. Subsequently, research on potential enzyme inhibitors is escalating broadly and is focused mostly on natural product derivatives such as peptides, polyphenolics, and terpenes. Numerous studies have investigated the therapeutic effects of various parts of *M. oleifera* plants in inhibiting key enzymes involved in the pathogenesis of non-communicable diseases which are leading causes of deaths worldwide namely hypertension, diabetes, hyperlipidemia, cardiovascular disease and kidney disorders (Mansurah et al.,

2015). Studies have revealed that Moringa leaf extract have enzyme inhibitory activities on α-glucosidase, pancreatic α-amylase, cholesterol esterase (Muhammad et al., 2016), proteases (Bijina et al., 2011), and angiotensin converting enzyme (Mansurah et al., 2015).

Hypertension is a multifactorial systemic chronic condition through macrovascular and microvascular changes at the structural and functional levels (Cunha et al., 2016). One in three adults worldwide has elevated blood pressure which causes around half of all deaths from stroke and heart disease (WHO, 2013) and it is predicted that the rate would increase by 60% in 2025 (Kearney et al., 2005). The pathogenesis of hypertension might be due to an increase activity of renin angiotensin aldosterone system (RAAS), kalikerenin kinin system and sympathetic nervous system. Amongst them over activation of RAAS is significant whereby angiotensin converting enzyme (ACE) plays a major role (Balasuriya and Rupasinghe, 2011). Accordingly, the World Health Organization and the International Society for Hypertension categorize ACE inhibitors as the first line of treatment along with diuretics and β-blockers (Mansurah et al., 2015).

ACE is classically associated with the renin-angiotensin system regulating blood pressure. Its physiological function is to convert angiotensin I to the potent vasopressor angiotensin II. Though ACE conversion of angiotensin I to angiotensin II is a normal regulatory process in the body, high ACE activity leads to increased concentration of angiotensin II and subsequently hypertension (Sharifi et al., 2013). Thus, inhibition of ACE is a promising way of controlling over expression of RAAS. Synthetic ACE inhibitors, such as captopril, lisinopril and enalapril, are widely used for the treatment of cardiovascular and renal disease. However, several side effects have been associated with the clinical use of ACE inhibitors. Consequently, search for novel and natural ACE inhibitors could greatly benefit hypertensive patients (Balasuriya and Rupasinghe, 2011).

A number of plants extracts have been recognized as *in vitro* ACE inhibitor whereby the effect has been ascribed to the host of flavonoid molecules (Park et al., 2003; Nyman et al., 1998). Additionally, flavonoids exhibited an ability to hinder different zinc metalloproteinases including ACE (Guerrero et al., 2012). Moreover, a study has demonstrated that niaziminin - a mustard oil glycoside initially isolated (along with other glycosides such as niazinin and niazimicin) from ethanolic extracts of *M. oleifera* leaves, at concentration of 1 mg and 3 mg/kg-body weight, caused a 16–22% and a 40–65% fall of mean arterial blood pressure on Wistar Rats (Faizi et al., 1992).

Similarly, ethanolic extract of *M. oleifera* pods has led to the isolation of thiocarbamate, isothiocyanate glycosides, methyl p-hydroxybenzoate and β-sitosterol which are known to have hypotensive effects (Toma and Deyno, 2014; Anwar and Bhanger, 2003). Also, the widespread combination of diuretic along with lipid and blood pressure lowering constituents make this plant highly worthwhile in cardiovascular disorders (Anwar et al., 2007). Moringa roots, leaves, flowers, gum and the aqueous infusion of seeds have been found to have diuretic action (Morton, 1991; Caceres et al., 1992) and such diuretic components are likely to play a complementary role in the overall blood pressure lowering effect of this plant. Additionally, another study has revealed that repeated oral administration of 3 mL/kg body weight aqueous extract of Moringa leaf /day for 6 weeks in spontaneously hypertensive rats, resulted in a significant decrease in systolic blood pressure compared to control (Kajihara et al., 2008).

1.5.4. Hypoglycemic Effects and Potential Anti-Diabetic Action of M. oleifera

Age, genetics, environment, and lifestyle influence the expansion of diabetes (Mbikay, 2012). Diabetes progression encompasses a complex network of interacting cellular and physiological modifications leading to insulin resistance, β cell failure leading to glucotoxicity- an excessive uptake of glucose by islet β cells (Robertson et al., 2004). This drives glycation reactions and the mitochondrial electron transport chain, producing reactive oxygen species (ROS), at levels beyond the antioxidation capacity of the cell. This oxidative stress impairs insulin synthesis and secretion, and initiates a cascade of cellular events that eventually lead to apoptosis and in turn Type 2 diabetes mellitus (Kaneto et al., 2007).

As a result, plant materials are constantly being scrutinized and explored for their potential usage as hypoglycemic agents (Efiong et al., 2013). Several recent studies have demonstrated that *M. oleifera* leaf extract has anti-diabetic properties (Gopalakrishnan et al., 2016). In a 21-day treatment in streptozotocin (STZ)-induced diabetes rats with the *M. oleifera* extract at a daily dose of 300 mg/kg-body weight, fasting plasma glucose levels and the post-prandial levels were reduced by 69% and 51%, respectively (Jaiswal et al., 2009). Likewise, another study has revealed that STZ- induced diabetes male rats treated with the low doses of Moringa seed powder caused a decreased in fasting blood glucose levels (Al-Malki and El-Rabey, 2015).

Similarly, it was found that treatment of STZ-induced hyperglycemic rats with aqueous leaf extract of *M. oleifera* at a dose of 200 mg/kg/body weight produced normoglycemia in 88.9% of animals under study by the end of the first week of treatment, and all the animals had their blood glucose brought to near normal by the end of second week. It was reported that the anti-diabetic activity of this miracle plant is primarily due to an array of antioxidant compounds such as glucomoringin, flavonoids, vitamin C and E (Al-Malki and El-Rabey, 2015; Adeeyo et al., 2013). Moreover, according to the literature, hypoglycemic effect was achieved through accentuation of insulin release from β cells of the islet of Langerhans of the pancreas, prevention of glucose uptake from gastrointestinal tract, and acting as α-glucosidase or pancreatic amylase enzyme inhibitors which prevent digestion of glucose and finally decreasing blood glucose levels (Adewole et al., 2007; Abdul Karim et al., 2005). Subsequently, it can be hypothesized that Moringa extract can increase antioxidant enzymes in the serum and thus, decreasing the level of ROS in the β-cells owing to the STZ induction (Mbikay, 2012).

1.5.5. Anti-Cancer Properties

The use of dietary and medicinal plant extracts as prophylactics against cancer is of utmost importance given the rise of cancer cases as well as the limited uses of the current therapies against the disease. Studies have revealed that Moringa can be used as an anti-proliferative agent, thereby impeding the growth of different types of cancer cells (Gopalakrishnan et al., 2016). Furthermore, researches have shown that this plant exhibits anti-cancer potential by interfering with the signal transduction cascade that stimulates cancer cell proliferation and progression (Tiloke et al., 2013). The inhibition of cancer cell proliferation is principally due to the presence of eugenol, a phenolic natural compound which targets E2F1/survivin in cancer cells, isopropyl isothiocynate and D-allose (Al-Sharif et al., 2013; Matsuda et al., 2007; Sui et al., 2005).

Moreover it was revealed that *M. oleifera* leaf extracts (0-300 µg/mL) induce apoptosis of human hepatocellular carcinoma cells (HepG2) by induction of caspase while downregulating the anti-apoptotic Bcl-xL protein, demonstrating that the extract inhibited cell proliferation (Jung et al., 2015). Additionally, boiled freeze-dried *M. oleifera* pods at 1.5%, 3.0% and 6.0%, respectively demonstrated inhibitory potential against azoxymethane-induced colon carcinogenesis and could therefore serve as a chemopreventive agent (Budda et al., 2011). Moringa aqueous leaf extract also increased the cytotoxicity of chemotherapy on pancreatic cancer (Panc-1) cells. Significant inhibition of Panc-1 cell survival was observed at an extract concentration of 0.75 mg/mL and the extract also reduced the overall expression of key NF-κB family proteins in the cells. Subsequently, the inhibition of the NF-κB signaling cascade, explains in part its attenuating ability of the Panc-1 cell survival (Berkovich et al., 2013). Recently, studies have been focused on the glycosylated isothiocyanate moringin (MG) or [4-(α-L-rhamnosyloxy) benzyl isothiocyanate; GMG-ITC], resulting from quantitative myrosinase-induced hydrolysis of glucomoringin (GMG) (4-(α-L-rhamnopyranosyloxy) benzyl GL), which is present in reasonable amounts in vegetables belonging to the Moringaceae family. MG has been reported to exert numerous beneficial activities, comprising protection against neurodegenerative disorders, anticancer, anti-inflammatory as well as antioxidant effects. A study has evaluated the potency of moringin on apoptosis induction and cell death in human astrocytoma grade IV CCF-STTG1 cells. Moringin revealed to be effective in inducing apoptosis through p53 and Bax activation and Bcl-2 inhibition. Concomitantly, *in vitro* results demonstrated antitumor efficacy of moringin in human malignant astrocytoma cells (Rajan et al., 2016).

Quinone reductase (QR) or NADPH:quinone oxidoreductase 1 (NQO1) is a phase II detoxifying enzyme which catalyzes the 2-electron reduction of a wide array of compounds especially quinones. The reduction of quinones to hydroquinones by QR, bypasses the formation of several carcinogenic compounds mainly semiquinone and also protects cells against ROS generated by quinones (Cuendet et al., 2006; Gerhäuser et al., 2003). Consequently, QR is widely used as the anti-carcinogenic phase II marker enzyme for assessing cancer chemopreventive agents (Kang and Pezzuto, 2004). Concurrently, in a study dichloromethnane M. oleifera extract was analyzed for its aptitude to induce QR activity on Hepa1c1c7 cells. It was revealed that the extract had an inductive effect on QR activity. The possible mechanism by which polyphenolic compounds induce QR gene expression is well understood. The induction of QR gene is regulated on the transcriptional level mediated by antioxidant response element (ARE), controlled by the nuclear factor E2-related factor 2 (Nrf2) (Nguyen et al., 2009) which is activated by polyphenols with antioxidant activity such as quercetin, kaempherol, glucosinolate and sulphoraphane. This mechanism up-regulates QR gene. In addition, glucosinolates (Forster et al., 2015), isothiocynates (Waterman et al., 2014) and glycosides (Faizi et al., 1992) present in different parts of Moringa plant could be responsible for QR induction as well as cancer cell anti-proliferation (Charoensin, 2014).

1.5.6. Alzheimer Disease and M. oleifera

With over 46 million people living with dementia globally and its anticipated rise to 131.5 million by 2050, this clinical condition also represents a major public health concern and a socio-economic burden. Alzheimer's disease (AD) is the most common neurodegenerative disorder and progressively compromises both memory and cognition, culminating in a state of full dependence and dementia, accounting for an estimated 60-80% of cases (Ansari and Khodagoli, 2013, Casey, 2012; World Alzheimer Report, 2015). Given the scale of this clinical condition potentiated by a global ageing population, it is vital to reflect on how natural resources such as bioactive foods, medicinal herbs and plant derived biofactors can be used to decrease the risks as well as manage the disease. With the current modern pharmacological therapy providing only short-term enhancement for AD, prophylactic contribution poses judicious course of action to relieve AD burden on the population.

Accumulated evidence suggests that naturally occurring phyto-compounds, such as polyphenolic antioxidants found in fruits, vegetables, herbs and nuts, may potentially hinder neurodegeneration and improve memory and cognitive function. Increasing epidemiological studies advocate that a healthy diet and nutrition might be a vital modifiable risk factor for AD (Hu et al., 2013). Thus, the mechanisms of action of bioactive compounds in parallel with the use of novel delivery techniques certainly provide the opportunity to explore these molecules as complementary to conventional medicine (Drever et al., 2009).

M. oleifera being a potential store house of antioxidants offers ample of health benefits to AD patients including modification of an array of brain monoamines namely dopamine, norepinephrine and serotonin (Obulesu and Rao, 2011). Ganguly and Guha (2008) proved that monoamines involved in memory loss can be altered by *M. oleifera* leaf extracts. The extracts have also been demonstrated to enhance memory via nootropics activity and provides considerable antioxidants such as vitamin C and E to combat oxidative stress in rat model of Alzeimer's Disease (Ganguly and Guha, 2008; Ganguly et al., 2005; Ganguly and Guha, 2006). Moreover, *M. oleifera* leaves extract significantly enhanced spatial memory and decreased neurodegeneration and oxidative stress markers like malondialdehyde (MDA) while increasing superoxide dismutase, catalase and acetylcholinesterase (AChE) activity in hippocampus in rats with age-related dementia. This mechanism might be due to the decreased oxidative stress and the enhanced cholinergic function (Sutalangka et al., 2013). Similarly, *M. oleifera* ethanolic extract exhibited moderate AChE inhibitory activity, inhibiting 47% of the latter activity (Mota et al., 2012).

1.5.7. Anti-Obesity Effect of M. oleifera

Obesity has been described as a disease due to an intricate interaction of genetic, environmental, endocrine, metabolic, cultural, and socioeconomic factors (Ahmed et al., 2014). Obesity is one of the leading causes of mortality worldwide and is commonly linked with dyslipidemia, coronary artery disease, hypertension, T2DM, non-alcoholic steatohepatitis, sleep apnea and certain types of cancers (Wooding and Rehman, 2014). Presently, two types of anti-obesity drugs are on the market namely orlistat (Xenical), which diminishes intestinal fat absorption through inhibition of pancreatic lipase (Drew et al., 2007) and sibutramine which is an anorectic (Tziomalos et al., 2009). However, due to their potentially perilous side effects, the potential of natural products (NPs) represent an outstanding alternative approach for developing future effective and safe anti-obesity therapies. A range of NPs, encompassing crude extracts and isolated bioactive compounds from plants, can induce weight reduction and prevent diet-induced obesity (Ahmed et al., 2014). Moreover, a growing body of evidence shows that NPs having anti-obesity effects can be organized into five categories based on their diverse mechanisms; namely (1) decreased lipid absorption, (2) decreased energy intake, (3) increased energy expenditure, (4) decreased pre-adipocyte differentiation and proliferation, or (5) decreased lipogenesis and increased lipolysis (Yun, 2010).

A study has evaluated the effectiveness of ethanolic extract of *M. oleifera* leaves in management of obesity induced by high cholesterol diet in rats. The results revealed that treatment with Moringa extract reduce food intake and BMI as well as ameliorate dyslipidemia, in obese rats. Serum leptin level showed significant decrease in obese groups and significant inhibition of serum MDA and nitric oxide levels was detected as a consequence of treatment with the extract. Furthermore, the treatment of obese group with the extract caused significant decrease in serum resistin level and a concomitant significant increase in serum adiponectin level. The reduction in BMI after Moringa treatment may be due to the presence of saponins and tannins which inhibits dietary lipid utilization (Ahmed et al., 2014). Likewise, another study demonstrated that *M. oleifera* extract has cholesterol reducing effect. Cholesterol homeostasis is controlled by the two processes, cholesterol biosynthesis in which HMG-Co-A reductase catalyzes rate limiting process and cholesterol absorption. The HMG-Co-A/mevalonate ratio has an inverse relationship to the activity HMG-Co-A reductase. It was found that the activity of this enzyme is significantly depressed by the extract (Jain et al., 2010).

Furthermore, the therapeutic potential of Moringa leaves on dyslipidemia induced in rabbits on a high-cholesterol (5%) diet (HCD) for 12 weeks was investigated. Treating these HCD rabbits with *M. oleifera* aqueous leaf extract, at the daily dose of 100 mg/kg-body weight for 12 weeks, has shown to reduce triglyceride and lipoprotein-cholesterol by about 50 and 75% respectively, and carotic plaque formation by 97% (Chumark et al., 2008). Additionally, β sitosterol is a plant sterol having similar structure to that of cholesterol, except for the substitution of an ethyl group at C24 of its side chain. It is understood that this compound has the aptitude to lower cholesterol by decreasing plasma concentrations of low density lipoprotein-cholesterol. Thus, β-sitosterol in the leaves of Moringa accounts for its hypolipidemic effect (Ahmed et al., 2014). Similarly, the presence of various natural antioxidants in the leaves of *M. oleifera* could thwart oxidation of low density lipoprotein-C with consequent increase in high density lipoprotein-C levels (Anwar et al., 2007; O'Byme et al., 2002). Therefore, these findings provide experimental evidence for the anti-obesity effect of *M. oleifera* extract.

CONCLUSION

The unprecedented worldwide rise in NCDs, as indicated by the World Health Organization, is a major cause of morbidity and mortality; consequently becoming an important health and economic burdens worldwide, including developing countries. Most NCDs, nevertheless, are multifactorial in nature, arising from complex interactions between the effects of multiple genes and environmental factors. Although modern pharmacotherapy has dwindled the mortality rate among the population, it has failed to stalk the rise. Additionally, expensive pharmacotherapy, side effects of treatment, limited new treatment coming on the market and drug resistance in other cases is adding on to this major challenge. Therefore, there is an urgent need to search for inexpensive preventive measures that may delay or abrogate the use of pharmacological agents to control this precipitous rise in the incidence of NCDs. Medicinal plants/ extracts/dietary food extracts can thus contribute in strengthening health care opportunities for people with NCDs as well as in the management of the biologic risk factors for NCDs at an early stage.

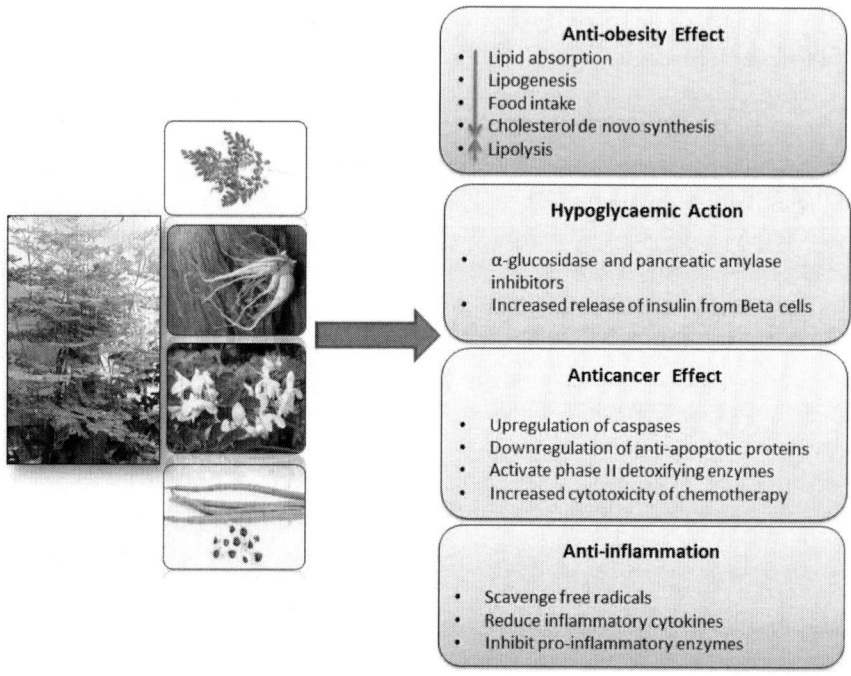

Figure 3. Summary of mechanism of action of *M. oleifera*.

M. oleifera tree has been revealed to possess a wide range of medicinal and therapeutic properties (Figure 3). Likewise, research has indicated the potential of different parts from Moringa plants to be used as a valuable source of bioactive compounds for development of functional food and nutraceuticals. These bioactives represent enormous promise as alternative approach to thwarting and managing NCDs due to their interaction with multiple targets involved in the onset of the disease. Understanding the nature of Moringa bioactive, their combinations and the molecular mechanism of action in eliciting bioactivity is also vital. Further epidemiological, follow-up studies and *in vitro* and *in vivo* evaluation are thus warranted in order to explore maximum therapeutic potential of the Moringa tree, for novel and effective utilization as pharmaceuticals and nutraceuticals for the management and/or prevention of NCDs. *M. oleifera* plant is certainly a miracle plant with massive potentials hitherto to be effusively explored in food application.

REFERENCES

Abdul karim, S.M., Long, K., Lai, O.M., Muhammad, S.K.S., Ghazali, H.M. (2005). Some physico-chemical properties of *Moringa oleifera* seed oil extracted using solvent and aqueous enzymatic methods. *Food Chem*.93: 253-263.

Abilgos, R. G., Barba, C. V. C. (1999). Utilization of Malunggay (*Moringa oleifera* Lam.) leaves in rice (Orya sativa L.) flat noodle-production. *Philippine J. Science*. 128:79-84.

Adeeyo, A. O., Adefule, A. K., Ofusori, D. A., Aderinola, A. A., Caxton-Martins, E.A. (2013). Antihyperglycemic effects of aqueous leaf extracts of mistletoe and *Moringa oleifera* in streptozotocin-induced diabetes wistar rats. *Diabetologia Croatica*. 42-3.

Adewole, S. O., Ojewole, J.A., Caxton-Martins, E.A. (2007). Protective effects of quercetin on the morphology of pancreatic beta cells of streptozotocin treated diabetic rats. *Afr J Tradit Compl Alternative Med*. 4:64-74.

Ahmed, H. H., Metwally, F. M., Rashad, H., Zaazaa, A. M., Ezzat, M., Salama, M. M. (2014). *Moringa oleifera* offers a Multi-Mechanistic Approach for Management of Obesity in Rats. *Int. J. Pharm. Sci. Rev. Res*. 29(2), 19: 98-106.

Al-Asmari, A. K., Albalawi, S. M., Athar, M. T., Khan, A.Q., Al-Shahrani, H., Islam, M. (2015). *Moringa oleifera* as an Anti-Cancer Agent against Breast and Colorectal Cancer Cell Lines. *See comment in PubMed Commons belowPLoS One*. 19:10(8):e0135814.

Al-Malki, A. L and El-Rabey, H. A. (2015). The anti-diabetic effect of low doses of *Moringa oleifera* Lam. seeds on streptozotocin induced diabetes and diabetic nephropathy in male rats. *Biomed. Res. Int*. 1-13.

Al-Sharif, I., Remmal, A., Aboussekhra, A. (2013). Eugenol triggers apoptosis in breast cancer cells through E2F1/survivin down-regulation. *BMC cancer*. 13(1):600.

Ansari, N., and Khodagholi, F. (2013). Natural Products as Promising Drug Candidates for the Treatment of Alzheimer's Disease: Molecular Mechanism Aspect. *Curr Neuropharmacol* 11:414–429.

Anwar, F and Bhanger, M.I. (2003). Analytical characterization of *Moringa oleifera* seed oil grown in temperate regions of Pakistan. *J Agric Food Chem* 51: 6558–6563.

Anwar, F., Latif, F., Ashraf, M., and Gilani, A.H. (2007). *Moringa oleifera*: A Food Plant with Multiple Medicinal Uses. *Phytother. Res*. 21:17–25.

Arora, D.S., Onsare, J.G., and Kaur, H. (2013). Bioprospecting of Moringa (Moringaceae): Microbiological Perspective. *Journal of Pharmacognosy and Phytochemistry*.http://search.proquest.com/assets/r20161.8.0.313.504/core/spacer.gif1(6):http://search.proquest.com/assets/r20161.8.0.313.504/core/spacer.gif 1193.

Atawodi, S.E., Atawodi, J.C., Idakwo, G.A., Pfundstein, B., Haubner, R., Wurtele, G., Bartsch, H., Owen, R.W. (2010). Evaluation of the polyphenol content and antioxidant properties of methanol extracts of the leaves, stem, and root barks of *Moringa oleifera* Lam. *J Med Food.* 13(3):710-6.

Bajpai, M., Pande, A., Tewari, S.K., Prakash, D. (2005). Phenolic contents and antioxidant activity of some food and medicinal plants. *International Journal of Food Sciences and Nutrition.* 56 (4): 287–291.

Balasuriya, B.W.N. and Rupasinghe, H.P.V. (2011). Plant flavonoids as angiotensin converting enzyme inhibitors in regulation of hypertension. *Functional Foods in Health and Disease.* 5:172-188

Balboa, E.M., Conde, E., Moure, A, Falque, E., Dominguez, H., (2013). *In vitro* antioxidant properties of crude extracts and compounds from brown algae. *Food chemistry.* 138:1764-1785.

Berkovich, L., Earon, G., Ron, I., Rimmon, A., Vexler, A., Lev-Ari, S. (2013). *Moringa oleifera* aqueous leaf extract down-regulates nuclear factor-kappaB and increases cytotoxic effect of chemotherapy in pancreatic cancer cells. *BMC Complement. Altern. Med.* 13: 212-219.

Bhatnagar, I and Kim, S.K. (2010). Marine antitumour Drugs: Status, shortfalls and strategies. *Mar Drugs* 8 (10): 2702-2720.

Bholah, K., Ramful-Baboolall, D., Neergheen-Bhujun, V.S. (2015). Antioxidant Activity of Polyphenolic Rich *Moringa oleifera* Lam. extracts in Food Systems. *Journal of Food Biochemistry.* 39(6).

Bijina, B., Chellappan, S., Krishna, J.G., Basheer, S.M.,Elyas, K.K., Bahkali, A.H., Chandrasekaran, M. (2011). Protease inhibitor from *Moringa oleifera* with potential for use as therapeutic drug and as seafood preservative. *Saudi Journal of Biological Sciences.* 18 (3): 273–281.

Bilto, Y. Y., Suboh, S., Aburjai, T., Abdalla, S. (2012). Structure-activity relationships regarding the antioxidant effects of the flavonoids on human erythrocytes. *Natural Science.* 4(9): 740-747.

Brunelli, D., Tavecchio, M., Falcioni, C., Frapolli, R., Erba, E., Iori, R., Rollin, P., Barillari, J., Manzotti, C., Morazzoni, P., D'Incalci, M. (2010). The isothiocyanate produced from glucomoringin inhibits NF-kB and reduces myeloma growth in nude mice *in vivo*. *Biochem. Pharmacol.* 79: 1141–1148.

Budda, S., Butryee, C., Tuntipopipat, S., Rungsipipat, A., Wangnaithum, S., Lee, J.S., Kupradinun, P. (2011). Suppressive effects of *Moringa oleifera* Lam pod against mouse colon carcinogenesis induced by azoxymethane and dextran sodium sulfate. *Asian Pac. J. Cancer Prev.* 12: 3221-3228.

Caceres, A., Saravia, A., Rizzo, S., Zabala, L., Leon, E.D., Nave, F. (1992). Pharmacologic properties of *Moringa oleifera*: 2: Screening for antispasmodic, anti-inflammatory and diuretic activity. *J Ethnopharmacol.* 36: 233–237.

Casey, G. (2012). Alzheimer's and other dementias. *Kai Tiaki Nursing New Zealand*, 18:20-24.

Cerf, M.E. (2013). Beta cell dysfunction and insulin resistance. *Front Endocrinol.* 4: 1-12.

Charoensin, S. (2014). Antioxidant and anticancer activities of *Moringa oleifera* leaves. *Journal of Medicinal Plant Research.* 8(7):318-325.

Chumark, P., Khunawat, P., Sanvarinda, Y., Phornchirasilp, S., Morales, N. P., Phivthong-Ngam, L., Ratanachamnong, P., Srisawat, S., Pongrapeeporn, K. U. (2008). The *in vitro* and *ex vivo* antioxidant properties, hypolipidaemic and antiatherosclerotic activities of water extract of *Moringa oleifera* Lam. leaves. *J. Ethnopharmacol.* 116: 439–446.

Crozier, A., Jaganath, I. and Clifford, M. N. (2006). In Plant Secondary metabolites. Occurance, structure and role in the human diet. A. Crozier, M.N. Clifford and H. Ashihara (eds) Blackwell Publishing Ltd. Oxford, UK. pp.1-22.

Cuendet, M., Oteham, C. P., Moon, R. C., Pezzuto, J. M. (2006). Quinone reductase induction as a biomarker for cancer chemoprevention. *J. Nat. Prod.* 69(3): 460-463.

Cunha, G. H., Fechine, F. Z., Frota Bezerra, F. A., Moraes, M. O., Silveira, E. R., Canuto, K. M., Moraes, M.E.A. (2016). Comparative study of the antihypertensive effects of hexane, chloroform and methanol fractions of essential oil of *Alpinia zerumbet in rats Wistar* Rev. *Bras. Pl. Med., Campinas*, 18 (1).113-124.

D'archivio, M., Filesi, C., Benedetto, R.D., Gargiulo, R., Giovannini, C and Masella, R. (2007). Polyphenols, dietary sources and bioavailability. *Ann Ist Super Sanita.* 43(4):348-361.

Dongmeza, E., Siddhuraju, P., Francis, G., Becker, K. (2006). Effects of dehydrated methanol extracted Moringa (*Moringa oleifera* Lam.) leaves and three of its fractions on growth performance and feed nutrient assimilation in Nile tilapia (Oreochromis niloticus (L.)). *Aquaculture.* 261 (1): 407-422.

Doughari, J. H., Human, I.S, Bennade, S. & Ndakidemi, P.A. (2009). Phytochemicals as chemotherapeutic agents and antioxidants: Possible solution to the control of antibiotic resistant verocytotoxin producing bacteria. *Journal of Medicinal Plants Research.* 3(11): 839-848.

Drever, B. D., Anderson, W. G., Riedel, G., Kim, D. H., Ryu, J. H., Choi, D. Y., Platt, B. (2009). The seed extract of *Cassia obtusifolia* offers neuroprotection to mouse hippocampal cultures. *J. Pharmacol. Sci.* 107:380–39210.

Drew, B. S., Dixon, A. F., Dixon, J. B. (2007). Obesity management: Update on orlistat. *Vasc Health Risk Manag.* 3: 817–821.

Efiong, E. E., Igile, G. O. Mgbeje, B. I. A., Out, E. A. and Ebong, P. E. (2013). Hepatoprotective and anti-diabetic effect of combined extracts of *Moringa oleifera* and *Vernonia amygdalina* in streptozotocin-induced diabetic albino Wistar rats. *Journal of Diabetes and Endocrinology.* 4(4): 45-50.

Fahey, J. W. (2005). *Moringa oleifera*: A review of the medicinal evidence for its nutritional, therapeutic, and prophylactic properties. Part 1. Trees Life J, 1, 5.

Faizi S., Siddiqui B. S., Saleem R., Siddiqui S., Aftab K. (1992). Isolation and structure elucidation of novel hypotensive agents, niazinin A, niazinin B, niazimicin and niaziminin A + B from *Moringa oleifera*: the first naturally occurring thiocarbamates. *J. Chem. Soc.* 1: 3237–3241.

Farooq Anwar, Sajid Latif, Muhammad Ashraf and Anwarul Hassan Gilani. *Moringa oleifera*: A Food Plant with Multiple Medicinal Uses. Phytotherapy Research, 21, 17–25, 2007.

Förster, N., Ulrichs, C., Schreiner, M., Müller, C. T., Mewis, I. (2014). Development of a reliable extraction and quantification method for glucosinolates in *Moringa oleifera. Food Chem.* 166:456-64.

Gagliano, N., Grizz,i F., Annoni, G. (2007). Mechanism of aging and liver functions. *Dig. Dis. Sci.* 25:118-123.

Ganguly, R. and Guha, D. (2008). "Alteration of brain monoamines & EEG wave pattern in rat model of Alzheimer's disease & protection by *Moringa oleifera*,". *Indian Journal of Medical Research.* 128 (6): 744–751.

Ganguly, R., and Guha, D. (2006). Protective role of an Indian herb, *Moringa oleifera* in memory impairment by high altitude hypoxic exposure: Possible role of monoamines. *Biogenic Amines.* 20:121–33.

Ganguly, R., Hazra, R., Ray, K., and Guha, D. (2005). "Effect of *Moringa oleifera* in experimental model of Alzheimer's disease: role of antioxidants,". *Annals of Neurosciences,* 12: 36–39.

Gerhäuser, C., Klimo, K., Heiss, E., Neumann, I., Gamal-Eldeen, A., Knauft, J., Liu, G.Y., Sitthimonchai, S., Frank, N. (2003). Mechanism-based in vitro screening of potential cancer chemopreventive agents. *Mutat. Res.* 523-524:163-172.

Ghosh, N. (2013). Anticancer effect of moringa oleifera leaf extract on human breast cancer cell, 1-57.

Gopalakrishnan, L., Doriya, K., Kumar, D. S. (2016). *Moringa oleifera*: A review on nutritive importance and its medicinal application. *Food Science and Human Wellness.* 5 (2):49–56.

Guerrero, L., Castillo, J., Quiñones, M., Garcia-Vallvé, S., Arola, L., Pujadas, G., and Muguerza, B. (2012). Inhibition of Angiotensin-Converting Enzyme Activity by Flavonoids: Structure-Activity Relationship Studies. *PLoS One.* 7(11): e49493.

Hassan, F., Ibrahim, M., (2013). *Moringa oleifera*: Nature is most nutritious and multi-purpose tree. *Int J Sci Res Pub.* 3(4): 1-5.

Heim, K. E., Tagliaferro, A. R., Bobilya, D. J. (2002). Flavonoid antioxidants: chemistry, metabolism and structure-activity relationships. *J Nutr Biochem.* 13(10):572-584.

Holt, E. M., Steffen, L. M., Moran, A., Basu, S., Steinberger, J., Ross, J. A., Hong, C., Sinaiko, A.R. (2009). Fruit and vegetable consumption and its relation to markers of inflammation and oxidative stress in adolescents. *J Am Diet Assoc.* 109(3): 414–421.

Hu, N., Yu, j., Tan, L., Wang, Y., Sun, L., Tan, L. (2013). Nutrition and the Risk of Alzheimer's Disease. *BioMed Research International* 13: Article ID 524820, 1-12.

Imdadul, H. S., Midul, H., Mostofa, S., Nazmuzzaman, N., Akhter, B., Rezuanul, I., Mohammad, A.H., Mostofa, J. (2016). A Review of the phytochemical and pharmacological profile of *Moringa oleifera* Lam. *Journal of Life Science and Biotechnology.* 3:75-87.

Iqbal, S and Bhanger, M. (2006). Effect of season and production location on antioxidant activity of *Moringa oleifera* leaves grown in Pakistan. *J Food Comp Anal.* 19:544–551.

Jain, P. G., Patil, S. D., Haswani, N. G., Girase, M. V., Surana, S. J. (2010). Hypolipidemic activity of *Moringa oleifera* Lam., Moringaceae, on high fat diet induced hyperlipidemia in albino rats. *Brazilian Journal of Pharmacognosy.* 20(6): 969-973.

Jaiswal, D., Kumar, Rai, P., Kumar, A., Mehta, S., Watal, G. (2009). Effect of *Moringa oleifera* Lam. leaves aqueous extract therapy on hyperglycemic rats. *J. Ethnopharmacol.* 123:392–396.

Josephine N. Kasolo, Gabriel S. Bimenya, Lonzy Ojok, Joseph Ochieng and Jasper W. Ogwal-Okeng. Phytochemicals and uses of *Moringa oleifera* leaves in Ugandan rural communities. Journal of Medicinal Plants Research Vol. 4(9), pp. 753-757, 4 May, 2010.

Jung, I. L., Lee, J. H., Kang, S. C. (2015). A potential oral anticancer drug candidate, *Moringa oleifera* leaf extract, induces the apoptosis of human hepatocellular carcinoma cells. *Oncol Lett.* 10(3): 1597–1604.

Kajihara, R., Nakatsu, S., Shiono, T., Shibata, I., Ishihara, M., Sakamoto, K., and Muto, N., Shokuhin, N., Kaishi, K.K. (2008). Antihypertensive effect of Water Extracts from Leaves of *Moringa oleifera* Lam. on Spontaneously Hypertensive Rats. *Journal of the Japanese Society for Food Science and Technology.* 55 (4): 183-185.

Kaneto, H., Katakami, N., Kawamori, D., Miyatsuka, T., Sakamoto, K., Matsuoka, T.A., Matsuhisa, M., Yamasaki, Y. (2007). Involvement of oxidative stress in the pathogenesis of diabetes. *Antioxid Redox Signal.* 9(3):355-66.

Kang, Y. H., Pezzuto, J. M. (2004). Induction of quinone reductase as a primary screen for natural product anticarcinogens. *Methods Enzymol.* 382:380-414.

Karim, N. A. A., Ibrahim, M. D., Kntayya, S. B., Rukayadi, Y., Hamid, H. A., Razis, A.F.A. (2016). *Moringa oleifera* Lam: Targeting Chemoprevention. *Asian Pacific Journal of Cancer Prevention.*17: 3675- 3686.

Kasolo, J. N., Bimenya, G.S., Ojok, L., Ochleng, J., Ogwal-Okeng, J. W. (2010). Phyochemicals and uses of *Moringa oleifera* leaves in Ugandan rural communities. *J. Med. Plant Res.* 4:753–757.

Kasolo, J. N.; Bimenya, G. S.; Okwi, A. L.; Othieno, E. M.; Ogwal-Okeng, J.W. (2011). Acute toxicity evaluation of *Moringa oleifera* leaves extracts of ethanol and water in mice. *Africa Journal of Animal and Biomedical Sciences.* 6(1), 40-44.

Kearney, P. M., Whelton, M., Reynolds, K., Muntner, P., Whelton, P.K. & He, J. (2005). Global burden of hypertension: analysis of worldwide data. *Lancet*, 365: 217-223.

Khawaja, T. M., Tahira, M., Mugal., and Ikram, U. H. (2010). *Moringa oleifera*: a natural gift-A review. *J. Pharm. Sci. & Res.* 2 (11): 775-781.

Koehn, F. E and Carter, G. T. (2005). The evolving role of natural products in drug discovery. *Nat Rev Drug Discov.* 4 (3):206-220.

Kris-Etherton, P. M., Hecker, K. D., Bonanome, A., Coval, S. M., Binkoski, A. E., Hilpert, K.., Griel, A. E., Etherton, T.D. (2002). Bioactive compounds in foods: their role in the prevention of cardiovascular disease and cancer. *Am. J. Med.* 113: 71S–88S.

Kumar, G. P., and Khanum, F. (2012). Neuroprotective potential of phytochemicals. *Pharmacogn Rev.* 6:81–90.

Kumar, P. S., Mishra, D., Ghosh, G., Panda, G. S. (2010). Medicinal uses and pharmacological properties of *Moringa oleifera. Int. J. Phytomed.* 2:210–216.

Lako, J., Trenerry, V. C., Wahlqvist, M., Wattanapenpaiboon, N., Sotheeswaran, S., Premier, R. (2007). Phytochemical flavonols, carotenoids and the antioxidant properties of a wide selection of Fijian fruit, vegetables and other readily available foods. *Food Chem.* 101:1727–1741.

Lee, Y. M., Seon, M. R., Cho, H. J., Kin, J., Park, H. J. Y. (2009). Benzyl isothiocyanate exhibits anti-inflammatory effects in murine macrophages and in mouse skin. *J Mol Med.* 87:1251–126.24.

Li, D., and Sinclair, A. J. (2002). Macronutrient innovations: the role of fats and sterols in human health. *Asia Pac J Clin Nutr.* 11: 155–62.

Limón-Pacheco, J and Gonsebatt, M. E. (2009). The role of antioxidants and antioxidant-related enzymes in protective responses to environmentally induced oxidative stress. *Mutat Res.* 674(1-2):137-47.

Lockett, C. T., Calvert, C. C., Grivetti L. E. (2002). Energy and micronutrient composition of dietary and medicinal wild plants consumed during drought. Study of rural Fulani, northeastern Nigeria. *Int. J. Food Sci. Nutr.* 51:195-208.

Mahajan, S. G., Mali, R.G., Mehta, A. A. (2007). Protective Effect of Ethanolic Extract of Seeds of *Moringa oleifera* Lam. Against Inflammation Associated with Development of Arthritis in Rats. *J Immunotoxicol.* 4(1): 39-47.

Mahomoodally, M. F., Mootoosamy, A., Wambugu, S. (2016). Traditional Therapies Used to Manage Diabetes and Related Complications in Mauritius: A Comparative Ethnoreligious Study. *Evidence-Based Complementary and Alternative Medicine* 4523828:1-25.

Manson, M. M. (2003). Cancer prevention – the potential for diet to modulate molecular signalling. *Trends in Molecular Medicine.* 9: 11–18.

Mansurah, A. A., Chioma, N.T., Ismaila, M., Abdulmalik, S.A., Williams, C., Wudil, A. M. (2015). Partial-purification and Characterization of Angiotensin Converting Enzyme Inhibitory Proteins from the Leaves and

Seeds of *Moringa oleifera*. *International Journal of Biochemistry Research & Review* 5(1): 39-48.

Matsuda, H., Ochi, M., Nagatomo, A., Yoshikawa, M. (2007). Effects of allyl isothiocyanate from horseradish on several experimental gastric lesions in rats. *European journal of pharmacology.* 561(1):172–81.

Mbikay, M. (2012). Therapeutic Potential of *Moringa oleifera* Leaves in Chronic Hyperglycemia and Dyslipidemia: A Review. *Front Pharmacol.* 3: 24.

Mohanty, S. K., Swamy, M. K., Middha, S. K., Prakash, L., Subbanarashiman, B., and Maniyam, A. (2015). Analgesic, Anti- inflammatory, Anti-lipoxygenase Activity and Characterization of Three Bioactive Compounds in the Most Active Fraction of *Leptadenia reticulata* (Retz.) Wight & Arn. – A Valuable Medicinal Plant. *Iran J Pharm Res.* 14(3): 933–942.

Molinari, G. (2009). Natural products in drug discovery: present status and perspectives. *Adv Exp Med Biol* 655: 13-27.

Morton, J. F. (1991). The horseradish tree, *Moringa pterigosperma* (Moringaceae). A boon to arid lands. *Econ Bot* 45: 318–333.

Mota, W. M., Barros, M. L., Cunha, P.E.L., Santana, M.V.A.; Stevam, C.S., Leopoldo, P.T.G., Fernandes, R.P.M. (2012). Evaluation of acetylcholinesterase inhibition by extracts from medicinal plants. *Rev. bras. plantas med.* 14(4).

Moyo, B., Masika, P. J., Hugo, A., Muchenje, V. (2011). Nutritional characterization of moringa (*Moringa oleifera* Lam.) leaves. *Afr. J. Biotechnology.* 10(60): 12925-12933.

Muhammad, H. I., Asmawi, M. Z., Karim Khan, N. A. (2016). A review on promising phytochemical, nutritional and glycemic control studies on *Moringa oleifera* Lam. in tropical and sub-tropical regions. *Asian Pacific Journal of Tropical Biomedicine.* 6 (10): 896–902.

Narayana, R. K., Sripal Reddy, K., Chaluvadi, M. R., Krishna, D. R. (2001). Bioflavonoids Classification, Pharmacological, Biochemical Effects and Therapeutic Potential. *Indian J of Pharmacol.* 33: 2-1.

Nazim, K., Chan Sun, M and Neergheen-Bhujun, V.S. (2015). Perceived health effects of *Moringa oleifera* Lam. following consumption in Mauritius. *Arch Med Biomed Res.*2:3-109.

Ndhlala, A. R., Mulaudzi, R., Ncube, B., Abdelgadir, H. A., Plooy, C. P., and Staden, J. V. (2014). Antioxidant, Antimicrobial and Phytochemical Variations in Thirteen *Moringa oleifera* Lam. Cultivars. *Molecules.* 19:10480-10494.

Ndiaye, M., Dieye, A.M., Mariko, F., Tall, A., Sall Diallo, A., Faye, B. (2002). Contribution to the study of the anti-inflammatory activity of *Moringa oleifera* (Moringaceae)]. *Dakar Med.* 47(2): 210-2.

Ng, V.W.L., Chan, J.M.W., Sardon, H., Ono, R.J., Garcia, J.M., Yang, Y.Y., Hedrick, J.L. (2014). Antimicrobial hydrogels: A new weapon in the arsenal against multidrug-resistant infections. *Advanced Drug Delivery Reviews.* 78: 46-62.

Nguyen, T., Nioi, P., Pickett, C.B. (2009). The Nrf2-antioxidant response element signaling pathway and its activation by oxidative stress. *J. Biol. Chem.* 284(20):13291-13295.

Nyman, U., Joshi, P., Madsen, L. B., Pedersen, T.B., Pinstrup, M., Rajasekharan, S., George, V., Pushpangadan, P. (1998). Ethnomedical information and in vitro screening for angiotensin-converting enzyme inhibition of plants utilized as traditional medicines in Gujarat, Rajasthan and Kerala (India). *J Ethnopharmacol* 60: 247–263.

O'Byme, D. J., Devaraj, S., Grundy, S. M., Jialal, I. (2002). Comparison of antioxidant effects of Concord grape juice flavonoids and αtocopherol on markers of oxidative stress in healthy adults. *Am J Clin Nutr.* 76: 1367–1374.

Obulesu, M., and Rao, D. M. (2011). Effect of plant extracts on Alzheimer's disease: An insight into therapeutic avenues. *J Neurosci Rural Pract.* 2(1): 56–61.

Olaiya, C. O., Soetan, K. O., and Esan A. M. (2016). The role of nutraceuticals, functional foods and value added food products in the prevention and treatment of chronic diseases. *African Journal of Food Science.* 10(10):185-193.

Park, P. J., Je, J. Y., Kim, S.K. (2003). Angiotensin I converting enzyme (ACE) inhibitory activity of hetero-chitooligosaccharides prepared from partially different deacetylated chitosans. *J Agric Food Chem.* 51: 4930–4934.

Patel, P., Patel, N., Patel, D., Desai, S., Meshram, D. (2014). Phytochemical analysis and antifungal activity of *Moringa oleifera. International Journal of Pharmacy and Pharmaceutical Sciences.* 6 (5) ISSN- 0975-1491.

Poston, W. S., and Foreyt, J. P. (2004). Sibutramine and the management of obesity. *Expert Opin Pharmacother.* 5: 633–642.

Raj, N. K, Sripal Reddy, K., Chaluvadi, M.R., Krishna, D.R. (2001). Bioflavonoids Classification, Pharmacological, Biochemical Effects and Therapeutic Potential. *Indian J of Pharmacol.* 33: 2-1.

Rajan, T.S., De Nicola, G.R., Iori, R., Rollin, P., Bramanti, P., Mazzon, E. (2016). Anticancer activity of glucomoringin isothiocyanate in human malignant astrocytoma cells. *Fitoterapia* 110: 1–7.

Ravipati, A. S., Zhang, L., Koyyalamudi, S. R., Jeong, S. C., Reddy, N., Bartlett, J., Smith, P. T., Shanmugam, K., Münch, G.W., Ming Jie, S., Manavalan and Vysetti, B. (2012). Antioxidant and anti-inflammatory activities of selected Chinese medicinal plants and their relation with antioxidant content. *BMC Complement. Altern. Med.* 12:1–14.

Razis, A. F., Ibrahim, M. D., Kntayya, S. B. (2014). Health benefits of *Moringa oleifera. Asian Pacific journal of cancer prevention.* 15:8571-8576.

Robertson, R. P., Harmon, J., Tran, P. O., Poitout, V. (2004). Beta-cell glucose toxicity, lipotoxicity, and chronic oxidative stress in type 2 diabetes. *Diabetes* 53(Suppl. 1), S119–S124.

Sáncheza, N. R., Spörndly, E., Ledin, I. (2006). Effect of feeding different levels of foliage of *Moringa oleifera* to creole dairy cows on intake, digestibility, milk production and composition. *Livestock Sci.* 101 (1-3): 24-31.

Sarwatt, S. V., Kapange, S. S and Kakengi, A. M. V. (2002). The effects on intake, digestibility and growth of goats when sunflower seed seed cake is replaced with *Moringa oleifera* leaves in supplements fed with *Chloris gayana* hay. *Agroforestry systems* 56:241-247.

Sashidhara, K. V., Rosaiah, J. N., Tyagi, E., Shukla, R., Raghubir, R., S. M. (2009). Rare dipeptide and urea derivatives from roots of *Moringa oleifera* as potential anti-inflammatory and antinociceptive agents. *Eur. J. Med. Chem.* 44: 432–436.

Shahbudin, S., Deny, S., Zakirun, A.M.T., Haziyamin, T.A.H., Akbar, J. B., and Taher, M. (2011). Antioxidant Properties of Soft Coral *Dendronephthya* sp. *International Journal of Pharmacology.*7: 263-267.

Sharifi, N., Souri, E., Ziai, S.A., Amin, G., and Amanlou, M. (2013). Discovery of new angiotensin converting enzyme (ACE) inhibitors from medicinal plants to treat hypertension using an *in vitro* assay. *Daru.* 21(1): 74.

Silva, M.F., Nishi, L., Farooqi, A., Bergamasco. R. (2014). The many health benefits of *Moringa oleifera. Journal of Medical and Pharmaceutical Innovation.* 1 (3): 9-12.

Soliva, C.R., Kreuzer, M., Foidl, N., Foidl, G., Machmuller, A. and Hess, H.D. (2005). Feeding value of whole and extracted Moringa oleifera leaves for ruminants and their effects on ruminal fermentation *in vitro*. *Animal Feed Science and Technology* 118 (1-2):47-62.

Sreelatha, S and Padma, P.R. (2009). Antioxidant activity and total phenolic content of *Moringa oleifera* leaves in two stages of maturity. *Plant Foods Hum Nutr.* 64(4):303-11.

Sui, L., Dong, Y., Watanabe, Y., Yamaguchi, F., Hatano, N., Tsukamoto, I., Izumori, K., Tokuda, M. (2005). The inhibitory effect and possible mechanisms of D-allose on cancer cell proliferation. *International journal of oncology.* 27(4):907–12.

Surh, Y. J. (2003). Cancer chemoprevention with dietary phytochemicals. *Nat. Rev. Cancer.* 3: 768–780.

Surh, Y. J. (2003). Cancer chemoprevention with dietary phytochemicals. *Natural Reviews in Cancer.* 3: 768–780.

Sutalangka, C., Wattanathorn, J., Muchimapura, S., and Thukham-mee, W. (2013). *Moringa oleifera* Mitigates Memory Impairment and Neurodegeneration in Animal Model of Age-Related Dementia. *Oxidative Medicine and Cellular Longevity.* Volume 2013, Article ID 695936, 9 pages.

Tejashree S. Masurekar, Vilasrao Kadam, Varsha Jadhav. (2015). Roles of *Moringa Oleifera* in Medicine - A Review. *World Journal of Pharmacy and Pharmaceutical Sciences.* 4(1):375-385.

Thurairajah, P. H., Syn, W. K., Neil, D. A., Stell, D., Haydon, G. (2005). Orlistat (xenical)-induced subacute liver failure. *Eur J Gastroenterol Hepatol.* 17: 1437–1438.

Tian Y, Zeng Y, Zhang J, Yang C, Yan L, Wang X, Shi C, Xie J, Dai T, Peng L, Zeng Huan Y, Xu A, Huang Y, Zhang J, Ma X, Dong Y, Hao S, Sheng J. High quality reference genome of drumstick tree (*Moringa oleifera* Lam.), a potential perennial crop. *Sci China Life Sci.* 2015; 58(7):627-38.

Tiloke, C., Phulukdaree, A., Chuturgoon, A. A. (2013). The antiproliferative effect of *Moringa oleifera* crude aqueous leaf extract on cancerous human alveolar epithelial cells. *BMC complementary and alternative medicine.* 13(1):226.

Toma, A and Deyno, S. (2014). Phytochemistry and pharmacological activities of *Moringa oleifera*. *International Journal of Pharmacognosy.* 1(4): 222-231.

Tsao, R. (2010). Chemistry and biochemistry of dietary polyphenols. *Nutrients.* 2:1231-46.

Tziomalos, K., Krassas, G.E., Tzotzas, T.(2009). The use of sibutramine in the management of obesity and related disorders: an update. *Vasc Health RisManag.* 5: 441–452.

Valko, M., Rhodes, C.J., Moncol, J., Izakovic, M., & Mazur, M. (2006). Free radicals, metals and antioxidants in oxidative stress-induced cancer. *Chemico-biology Interactions.* 160(1):1-40.

Waterman, C., Cheng, D. M., Rojas-Silva, P., Poulev, A., Dreifus, J., Lila, M. A., Raskin, I. (2014). Stable, water extractable isothiocyanates from *Moringa oleifera* leaves attenuate inflammation *in vitro. Phytochemistry.* 103:114–122.

Watzl, B. (2008). Anti-inflammatory effects of plant-based foods and of their constituents. *Int J Vitam Nutr Res.* 78(6):293-8.

WHO Global Health Observatory Data Repository [online database]. Geneva, World Health Organization, 2013 (http://apps.who.int/gho/data/view. main, accessed 21 May 2013).

Wiernsperger, N. F. (2003). Oxidative stress as a therapeutic target in diabetes: revisiting the controversy. *Diabetes Metabolism* 29: 579-585.

Wooding, A. E., Rehman, I. (2014). Obesity and prostate cancer: Is there a link. *e-SPEN Journal,* 9: e123ee130.

World Alzheimer Report 2015. http://www.worldalzreport2015.org/ (accessed 29 April 2014).

World Health Organization (2013). A global brief on hypertension. Silent killer, global public health crisis. World Health Day 2013, 40.

World Health Organization (WHO). Diet, Nutrition and the Prevention of Chronic Diseases: Report of a Joint WHO/FAO Expert Consultation. WHO Technical Report Series 916. WHO, Geneva, 2003.

World Health Organization (WHO). Global status report on non-communicable diseases. 2014.

Yang, T., Yan, Z., Jing, Z., ChengGuang, Y.,Liang, Y., XuanJun, W., ChongYing, S., Jing, X., TianYi, D., Lei, P., Yu Zeng, H., AnNi, X., YeWei, H., JiaJin, Z., Xiao, M., Yang, D., ShuMei, H., Jun, S. (2015). High quality reference genome of drumstick tree (*Moringa oleifera* Lam.), a potential perennial crop. *Science China Life Sciences.* 58 (7): 627–638.

Yun, J. W. (2010). Possible anti-obesity therapeutics from nature – A review. *Phytochemistry.* 71: 1625–1641.

Yunus, S. M., Fayazuddin, M., Ahmad, F., Kumar, A. (2013). An experimental evaluation of anti-inflammatory activity of *Moringa oleifera* seeds. *International Journal of Pharmacy and Pharmaceutical Sciences.* 5 (3).

Zaku, S. G., Emmanuel, S., Tukur, A. A., Kabir, A. (2015). *Moringa oleifera*: An underutilized tree in Nigeria with amazing versatility: A review. *African Journal of Food Science.* 9(9): 456-461.

In: Polyphenolics
Editor: Patricia Clark

ISBN: 978-1-53610-709-8
© 2017 Nova Science Publishers, Inc.

Chapter 4

IN VIVO ANTIOXIDANT AND ANTI-INFLAMMATORY EFFICACY OF POMEGRANATE POLYPHENOLS AND THEIR METABOLITES

Piteesha Ramlagan[1], Nawraj Rummun[1], Theeshan Bahorun[2], PhD, and Vidushi S. Neergheen-Bhujun[1,], PhD*

[1]Department of Health Sciences, Faculty of Science and ANDI Centre of Excellence for Biomedical and Biomaterials Research, MSIRI Building, University of Mauritius, Reduit, Republic of Mauritius
[2]ANDI Centre of Excellence for Biomedical and Biomaterials Research, MSIRI Building, University of Mauritius, Reduit, Republic of Mauritius

ABSTRACT

Punica granatum L. (pomegranate) has been valued for its medicinal properties since ancient times and is still being used in folk medicine across the world. The different anatomical parts of pomegranate have been professed with different health benefits in different indigenous cultures and associated traditional medicine. The dietary nature of pomegranate has attracted significant scientific interest at validating

---

* Corresponding Authors E0mail: v.neergheen@uom.ac.mu, tbahorun@uom.ac.mu.

its ethnomedicinal uses as well as promoting its use as a functional food to mitigate chronic human diseases. The prophylactic effects of pomegranate have been attributed to its polyphenolic richness, which has shown potent antioxidant and anti-inflammatory capacities, both *in vitro* and *in vivo*. Oxidative stress and high inflammation have been reported as the underlying pathophysiological hallmarks of a number of chronic diseases and attenuating the inflammatory state and increasing the antioxidant status constitute an important target in managing the latter. Dietary pomegranate polyphenols have been shown to reduce plasma oxidative stress markers namely protein carbonyls and malondialdehyde as well as to modulate C-reactive protein, sE-selectin, TNF-α and IL-6. The *in vivo* pharmacological and therapeutic effect of pomegranate polyphenol consumption is mostly attributed to the bio-transformed metabolites of pomegranate bioactive polyphenols, following ingestion in the human body, as evidenced by increasing amount of literature available from pomegranate pharmacokinetic studies. This chapter, therefore, reviews findings on the bio-activity of pomegranate polyphenols and their bio-transformed metabolites with emphasis on their *in vivo* anti-oxidative and anti-inflammatory mechanisms of action. The chapter will conclude on the relevance and potentials of pomegranate polyphenols in clinical trials.

Keywords: pomegranate polyphenols, antioxidant, anti-inflammatory

INTRODUCTION

Punica granatum L. (pomegranate), belonging to the Punicaceae family, is one of the oldest edible plants that have been used since ages for the treatment of many health problems (Patel et al., 2008; Çam et al., 2009; Daglia, 2012). Pomegranate is native to Iran but is now cultivated in several countries whereby at least 500 cultivars have evolved (Stover and Mercure, 2007). The long history of culinary uses of pomegranate fruit concurrent with the safe ethnomedicinal usage of both edible and non-edible parts of pomegranate (Table 1) in different regions of the world highlights its ability to be exploited as an effective and potentially non-toxic functional food. Furthermore, the polyphenolic richness of the plant parts provides an interesting reservoir of compounds which can act synergistically or in an additive effect to mediate prophylactic therapeutic effects. Because of these reasons, recent decades have witnessed an explosion in scientific investigation related to the health benefits of pomegranate, including epidemiological studies, *in vitro* structure elucidation and potential mechanism of action of bioactive compounds derived from pomegranate plant parts as well as clinical trials. Pomegranate fruit and juice are considered as antioxidant functional foods with anti-diabetic, anti-bacterial, anti-carcinogenic, anti-atherogenic, and anti-hypertensive potential amongst others (Table 1) (Jurenka, 2008).

Oxidative stress is defined as the over-production of reactive oxygen species (ROS); whilst the body possess an innate defence mechanism against ROS, decreased in elimination of the latter due to impairment of such system aggravate cellular redox balance. Oxidative stress state causes oxidative damage of macromolecules that in turn generate secondary reactive species leading to cellular dysfunction and thus disease development (Dalle-Donne, 2006) such as diabetes (Schaffer et al., 2012), atherosclerosis (Bonomini et al., 2008) and cancer (Reuter et al., 2010). For instance, peroxidation of membrane lipids due to ROS attack on polyunsaturated fatty acids on the membrane leads to decreased membrane fluidity, inactivation of membrane bound proteins and production of chemically reactive products such as malondialdehyde (MDA), 4-hydroxy-2-nonenal (HNE), 2-propenal (acrolein), glyoxal and 4-oxo-2-nonenal (ONE). Consecutively, some of these aldehydes react with DNA, proteins and phospholipids thereby altering the functionality of these molecules (Uchida, 2003; Dalle-Donne, 2006). Oxidative modifications of proteins lead to proteins dysfunction by altering protein structure, cleaving peptide backbone, forming cross-links and modifying side chains (Dalle-Donne, 2006). Cellular DNA damage induced by hydroxyl radical leads to modification of bases and sugars that have potential susceptibility for cancer development. As such, 8-hydroxy-2'-deoxyguanosine (8-OHdG) is a product of DNA oxidation to the base 2'-deoxyguanosine (Cooke et al., 2003). Furthermore, excess of ROS initiates inflammatory responses due to activation of the inflammatory transcription factor NFκB (Gloire et al., 2006). Literature data abounds in examples showing the high antioxidant potentials of pomegranate fruits and parts *in vitro*, attributed to polyphenolic richness mainly phenolics, hydrolyzable tannins, flavonoids and anthocyanins (Fischer et al., 2011; Zhao et al., 2014). These moieties confer specific structure-function properties that determine antioxidative capacities. As such the number of H-donating hydroxyl groups and compound configuration are important structural characteristics (Balasundram et al., 2006). For instance, gallic acid with a maximum of 3 hydroxyl groups has higher ability to donate hydrogen atom and stabilize radicals than other phenolics with one or two hydroxyls (Cuvelier et al., 1992). Polyphenols generally scavenge free radicals by acting as reducing agents and are themselves oxidized leading to the generation of new radicals that are stabilized by the resonance of the aromatic nucleus due to the delocalization of electrons (Cuvelier et al., 1992; Patel et al., 2010). Therefore, the larger number of aromatic rings in a compound, the greater would be its antioxidant potential. Furthermore, the metal ions scavenging activity of flavonoids has been attributed to the presence of at least one of the three structural domain as binding sites for metal ions in the flavonoid molecules, namely; (1) the catechol moiety in the ring B, (2) the 3-hydroxyl and 4-oxo groups in the heterocyclic ring C, and (3) the 4-oxo and 5-hydroxyl groups between the C and A rings (Malešev & Kuntić 2007; Procházková et al., 2011; Crascì et al., 2016). Flavonoids and hydrolyzable tannins richness in pomegranate provide an additional basis of antioxidant potency due to the presence of several aromatic rings and hydroxyl groups per molecule. As such, Gil et al. (2000) have shown that hydrolyzable tannin, phenolic acids and anthocyanin such as punicalagin, gallic/ellagic acid and cyaniding-3-glucoside respectively exhibited good antioxidant activities with punicalagin showing greater potentials.

Table 1. Ethnomedicinal uses of pomegranate fruit/parts

Pomegranate fruit/ part	Treatment	Reference
Ripe fruit	Brain disease, chest pain, colic, colitis, diarrhea, dysentery, earache, leucorrhea, menorrhagia, oxyuriasis, paralysis, rectocele	Schubert et al., 1999; Bagri et al., 2009
Pericarp	Atherosclerosis, cancer, colic, colitis, diabetes, diarrhea, dysentery, inflammation, leucorrhea, menorrhagia, oxyuriasis, paralysis, rectocele	Abdel-Hady, 2013; Schubert et al., 1999
Mesocarp	Atherosclerosis, biliousness, bronchitis, cancer, destruction of parasitic worms, diabetes, diarrhea, dysentery, piles, inflammation	Abdel-Hady, 2013; Bagri et al., 2009
Bark	Biliousness, bronchitis, colic, colitis, destruction of parasitic worms, diarrhea, dysentery, leucorrhea, menorrhagia, oxyuriasis, paralysis, piles, rectocele	Schubert et al., 1999; Bagri et al., 2009
Root	Colic, colitis, diarrhea, dysentery, leucorrhea, menorrhagia, oxyuriasis, paralysis, rectocele	Schubert et al., 1999
Flower	Bronchitis, diabetes, diarrhea, dysentery throat inflammation	Bagri et al., 2009

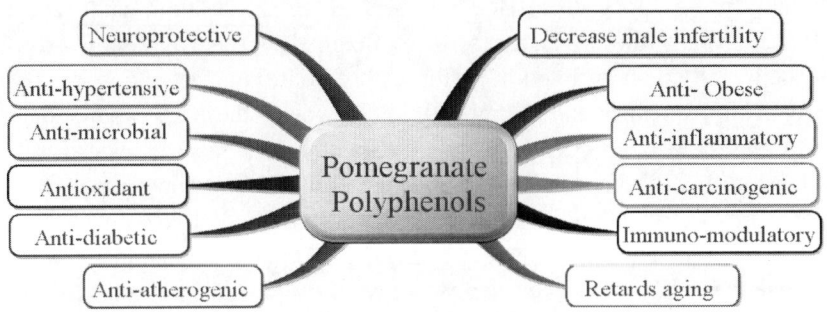

Figure 1. Bioactive effects of pomegranate phytochemicals *in vitro*.

Inflammation is a normal immune reaction in response to tissue injury or invading exogenous molecules/pathogens and can be either localized or systemic. Under physiological conditions, successful acute inflammation is followed by a resolution phase during which pro-inflammatory processes are suppressed and physiological homeostasis is achieved. However, an on-going inflammatory event or an impaired resolution phase characterized by continuous pro-inflammatory processes results in chronic inflammation (Maskrey et al., 2011; Fullerton and Gilroy 2016). During chronic inflammatory conditions, sustained production of ROS enhances expression of pro-inflammatory genes, via intra cellular signalling cascades, stimulating production of inflammatory mediators. The inflammatory mediators in turn causes migration of more inflammatory cells, therefore creating a vicious cycle of ROS production and inflammatory cell recruitment at the site of inflammation (Biswas, 2016). Chronic inflammation is implicated in the progression of multiple pathologies including autoimmune rheumatoid arthritis (Ghavipour et al., 2016), osteoarthritis (Robinson et al., 2016), hypertension (Solak et al., 2016), diabetes (Donath and Shoelson, 2011) and cancer (Mantovani et al., 2008) amongst others. Therefore modulating the pathogenic inflammatory pathways provides ways to mitigate the effect of such diseases. However, long term administration of classically used non-steroidal anti-inflammatory drugs (NSAIDs) is associated with severe adverse effects. Furthermore, NSAIDs only provide symptomatic relief and do not alter the progression of disease as shown in the case of rheumatoid arthritis (Quan et al., 2008). Consequently, recent advances in the field of biologics has led to novel anti-inflammatory therapies which make use of monoclonal antibodies to target key inflammatory mediators retarding disease progression (Haraoui and Bykerk, 2007). For instance, the efficacy of monoclonal antibodies directed against TNF-α (infliximab) has been clinically shown in the treatment of anklylosing spondylitis (Baraliakos et al., 2011; Elalouf and Elkayam, 2015). Similarly the effectiveness of tocilizumab, an anti-IL-6 antibody, has been clinically demonstrated in multiple chronic inflammatory diseases (Quan et al., 2008; Puchner and Blüml, 2015). However, the high production cost impacting the patients overall treatment expenditure, limits the widespread use of these drugs (Chames et al., 2009; Quan et al., 2008). Dietary polyphenols with antioxidative potential remain an opportune alternative to NSAIDs and biologics. The anti-inflammatory potential of plant flavonoids and other polyphenolics has been extensively reviewed by González et al., (2011) and Joseph et al., (2016). Structure activity relation studies revealed that presence of an unsaturated C-ring a carbonyl group at C-4 and the number and position of hydroxyl groups at ring B, are all important requirements for the anti-inflammatory activity of a flavonoid compound (Lago et al., 2014) and that glycosylation of the flavonoids tend to neutralise its anti-inflammatory potential (Lago et al., 2014). Polyphenol rich pomegranate extracts are used to mitigate disease condition as part of different traditional medicine across the world (Table 1) and has been shown to possess potent anti-oxidative and anti-inflammatory activities in both pre-clinical and clinical studies. In this chapter, we further highlight the promising antioxidant and anti-inflammatory effects of pomegranate polyphenols both in animal models and in human clinical trials.

PHYTOCHEMICAL DISTRIBUTION IN POMEGRANATE

The pomegranate tree is a deciduous shrub that can reach a height of 7 m. The fruit is grenade-shaped, crowned by a calyx and can reach up to 12 cm wide when ripe (Jurenka, 2008). Most parts of the pomegranate plant have been reported for their phytochemical content in relation to pharmacological effects.

Pericarp and Mesocarp

The pericarp (outer layer of the fruit) is yellowish brown with reddish mottling. The mesocarp (rind) is yellowish and occurs in thin curved pieces with impressions of the seeds (Duke and Ayensu, 1985). The pericarp and mesocarp (together termed as peel) constitute approximately 50% of the fruit's weight. The peel has more potent activities than other pomegranate parts due to its rich source of hydrolyzable tannins (gallagic acid, ellagic acid glycosides, punicalagin, punicalin, pedunculagin, corilagin, casuarinin, gallagyldilacton, tellimagrandin, granatin A, granatin B) in addition to phenolic acids (gallic, ellagic, caffeic, chlorogenic, p-coumaric, ferulic and vanillic acids), flavonoids (luteolin, quercetin, kaempferol, genistein), flavan-3-ols ((+) catechin, (-) epicatechin, catechin gallates), anthocyanidins (delphinidin, cyanidin, pelargonidin) and anthocyanins (pelargonidin-3-glucoside, pelargonidin-3,5-diglucoside, cyanidin-3-glucoside, cyanidin-rutinoside, cyanidin-pentoside, cyanidin-3,5-diglucoside, delphinidin 3,5-diglucoside, delphinidin 3-glucoside) (Viuda-Martos et al., 2010; Fischer et al., 2011; Abdel-Hady, 2013; Sreekumar et al., 2014; Lansky and Newman, 2007; Mohammad and Kashani, 2012, Ali et al., 2014). It is however noteworthy that anthocyanins are present in the pericarp but not mesocarp (Fischer et al., 2011) (Figure 2).

Aril

The arils are seeds that are surrounded by the small amount of tart, red juice. The seeds comprise 20% of the arils and 80% pulp (Afaq et al., 2005). The arils are separated by the mesocarp (Jurenka, 2008). The arils are rich in anthocyanins (pelargonidin-3-glucoside, pelargonidin-3,5-diglucoside, cyanidin-3-glucoside, cyanidin-rutinoside, cyanidin-pentoside, cyanidin-3,5-diglucoside, delphinidin 3,5- diglucoside, delphinidin 3-glucoside), phenolic acids (gallic acid, vanillic acid 4-glucoside, protocatechuic acid-derivative, caffeic acid hexoside and/or derivatives, coumaric and ferulic acids hexosides) and hydrolyzable tannins (ellagic acid and glycosides, digalloylhexoside, hexahydroxydiphenoyl (HHDP)-hex, bis-HHDP-hex, digalloyl HHDP-hex, punicalagin) (Jaiswal et al., 2010; Fischer et al., 2011) (figure 2).

Seed

Pomegranate seeds are rich in lignans (isolariciresinol, secoisolariciresinol, pinoresionol, medioresionol, syringaresinol), phenolic acids (ellagic acid, 3,3'-di-*O*-methylellagic acid, 3,3',4'-tri-*O*-methylellagic acid), conjugated fatty acids (punicic acid), non-conjugated fatty acids (linoleic, oleic, palmitic and stearic acids), sterols (daucosterol, camesterol, stigmasterol, β-sitosterol, cholesterol) and sex steroids (17-α-estradiol, estrone, testosterone, estriol) (Bonzanini et al., 2009; Lansky and Newman, 2011) (Figure 2).

Pulp

The pulp is rich in sugars (glucose, fructose, and sucrose), phenolic acids (gallic, ellagic, caffeic, chlorogenic and *p*-coumaric acids), flavonoids ((+) catechin, (-) epicatechin, epigallocatechin-3-gallate, quercetin, and rutin) anthocyanins (pelargonidin-3-glucoside, pelargonidin-3,5-diglucoside, cyanidin-3-glucoside, cyanidin-3,5-diglucoside, delphinidin 3,5- diglucoside, and delphinidin 3-glucoside) and lignans (isolariciresinol, secoisolariciresinol, pinoresinol, syringaresinol) (Bonzanini et al., 2009; Lansky et al., 2011) (Figure 2).

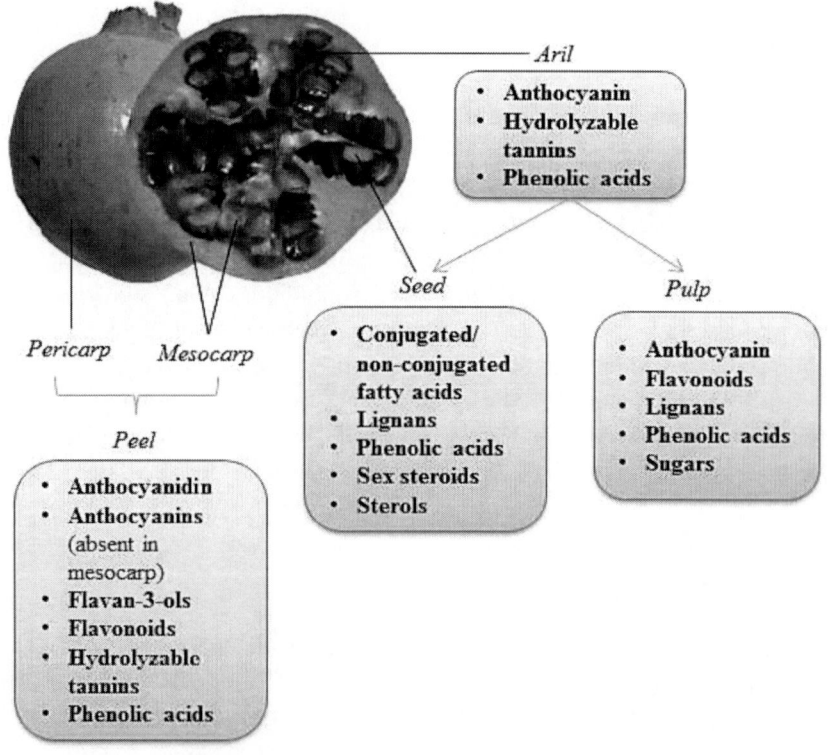

Figure 2. Selected compounds present in different parts of pomegranate.

MODULATORY EFFECT OF POMEGRANATE PHYTOCHEMICALS *IN VIVO*

Numerous studies have evaluated the prophylactic effects of pomegranate consumption against diabetes, obesity, hypertension, cancer. The disease mitigating effect of the extracts has been attributed to its potential to lower the patho-physiological concentration of inflammatory mediators to physiological level alongside improving the oxidative balance. Some of the outcomes of the investigations of pomegranate's effect in diseases underlying oxidative stress and inflammation will be discussed in the following sections.

Pain

Nociception, the perception of pain, is one of the cardinal features of an inflammatory response (Kidd and Urban, 2001). Sarker et al., (2012) used Swiss albino mice and Long-Evans rats model to study abdominal writhing, induced by intraperitoneal administration of acetic acid solution (0.7%) at a dose of 0.1 mL/10g body weight and reported the anti-nociceptive activity of pomegranate flower extracts, at an oral dose of 200 mg/kg body weight, to be more potent than diclofenac at an oral dose of 50 mg/kg body weight. In the same vein, using formalin test in male Swiss mice and male Wistar rats, González-Trujano et al., (2015) reported a significant dose dependent decrease in nociceptive response induced by pomegranate ellagitannins, with the anti-nociceptive response being evident from a dosage of 10 mg/kg body weight of pomegranate extract and complete inhibition of nociception at a dosage of 100 mg/kg body weight. González-Trujano et al., (2015) further reported that pomegranate extracts were more effective in decreasing the nociceptive response compared to diclofenac at a dosage of 100 mg/kg body weight. The decrease was consistent with both neurologic and inflammatory phase of nociception. The anti-nociceptive potential of pomegranate polyphenol has been reported by different group of investigators (Labib and El-Ahmady, 2015; Saad et al.,2014; Bensaad and Kim, 2015; Salwe and Sachdev, 2014), and therefore justify the use of pomegranate in arthritic pain management as part of the Italian folk medicine (Leporatti and Ghedira, 2009). A well-known adverse effect to commonly used analgesic including NSAIDs in the pain management is ulceration of the gastric-mucosa. Interestingly, intra-peritoneal administration of pomegranate extract (30 mg/kg body weight) exerted antinociceptive effect without producing any visible gastric lesion compared to indomethacin (20 mg/kg body weight), which produces severe gastric lesions (González-Trujano et al., 2015). Similar observations were made by Labib and El-Ahmady (2015) when reporting the anti-nociceptive activity of ellagic acid rich pomegranate peel extracts compared to indomethacin. Pomegranate ellagitannin rich extract further exerted significant protection of the gastric mucosa, against ethanol-induced gastric insults (González-Trujano et al., 2015).

Edema

Edema, a major feature of inflammation that is commonly investigated when assessing the anti-inflammatory activity in animal models was found to be effected by the pomegranate extracts. Labib and El-Ahmady (2015) reported the anti-edematogenic activity of orally administered ellagic acid rich pomegranate peel extract (at a dosage of 200 mg/kg body weight) in carrageenan-induced rat paw edema rat models, to be considerably more than that of orally administered indomethacin (at 20 mg/kg body weight) after 2 hours of edema induction. Similar inhibition of paw edema volume was observed in both Swiss albino mice and Long-Evans rats following prior administration of pomegranate flower extract (Sarker et al., 2012). Edema is known to be a result of altered vascular permeability and subsequent plasma leakage, in response to neutrophils; therefore inhibiting neutrophil recruitment at the inflammatory site reduces edema (Kenne et al., 2012). Substantial literature evidence now illustrates the involvement of pro-inflammatory cytokines TNF-α in promoting neutrophil migration (Marques et al., 2016) and altering vascular permeability (Sawant et al., 2014). TNF-α mediates remodelling of endothelial cells by disrupting the cell adherent junctions, thereby inducing cytoskeletal changes which increase the endothelial permeability (Sawant et al., 2014). Using a male Wister rat acute peritonitis model, Marques et al., (2016) reported pomegranate leaf extract (at a dosage of 250 mg/kg body weight) to inhibit the migration of neutrophils into the peritoneal cavity from the blood. The authors further investigated the level of circulating inflammatory biomarkers and reported pomegranate leaf extract treated rat to present a lower level of TNF-α in both serum and peritoneal cavity in comparison to vehicle treated rats. Pre-treatment of LPS-injected rats with pomegranate leaf extract resulted in a down-regulation of TNF-α level at both mRNA and protein levels (Marques et al., 2016). Polyphenol rich pomegranate extract exerts its anti-inflammatory potential by modulating inflammatory markers as evidenced from human clinical trials which corroborate the beneficial effect of polyphenols rich pomegranate juice against multiple diseases having inflammatory basis as part of their pathogenesis.

Atherosclerosis

Increased ROS plays a pivotal role in the pathogenesis of atherosclerosis as these reactive species, especially hydroxyl radicals, damage cell membranes and nuclei. ROS causes lipid peroxidation which leads to formation of oxidized low density lipoproteins (oxLDL), a mediator of atherosclerosis (Bonomini et al., 2008). The ROS scavenging potential of pomegranate polyphenols provides one of the major pomegranate anti-atherogenic properties.

Daily supplementation of 50 mL of pomegranate juice for two weeks to healthy volunteers resulted in a reduction in the susceptibility of plasma to 2,2'-azobis(2-amidinopropane) dihydrochloride (AAPH)-induced lipid peroxidation and an increase in plasma total antioxidant status. Pomegranate juice also showed an increase in serum paraoxonase (PON) and associated decrease in high density lipoprotein (HDL) and low density lipoprotein (LDL) oxidation induced by copper ion (Aviram et al., 2000). PON1 is a glycoprotein that when bound to HDL, has antioxidant property with lactone and arylesterase activities. PON1 confers anti-atherogenic properties by lowering oxidation of LDL and HDL, inhibiting oxLDL induced pro-inflammatory response and hydrolysing hydrogen peroxide (Mogarekar et al., 2016). Pomegranate juice also decreased plasma protein carbonyl and MDA levels while increasing the erythrocytes glutathione level in healthy volunteers after 15 days of consumption of pomegranate juice. It was noteworthy that 1 week after stopping pomegranate juice consumption, assessment of these biomarkers showed that the decrease in carbonyls and MDA was still maintained (Matthaiou et al., 2014), with ROS scavenging potentials of pomegranate polyphenols accountable for these changes. Increase in glutathione is attributable to the polyphenolic richness as quercetin, but not its conjugates, has been shown to increase transactivation of the enzyme γ-glutamylcysteine synthetase (γGCS) that is involved in synthesis of glutathione (Moskaug et al., 2005). Quercetin also activates nuclear factor-erythroid 2p45-related factor-2 (Nrf2) which binds to antioxidant responsive element (ARE) or electrophile responsive element (EpRE) leading to transcription of γGCS (Granado-Serrano et al., 2012; Moskaug et al., 2005).

Consumption of pomegranate juice decreased collagen-induced platelet aggregation (Aviram et al., 2000) most probably by its ability to scavenge free radicals leading to attenuation of platelet activation induced by oxidative stress (Freedman, 2008). *In vivo* studies in atherosclerotic apolipoprotein E-deficient (E^0) mice, which are under oxidative stress, showed that pomegranate increased serum total antioxidant status and decreased LDL oxidation induced by both copper ion and macrophages under oxidative stress (Aviram et al., 2000). As a consequence of lipid peroxidation in macrophages during oxidative stress, LDL undergoes oxidation (Fuhrman et al., 1994). Ox-LDL in turn generates foam cells from macrophages, accumulates cholesterol, affects coagulation pathways and stimulates inflammatory and thrombotic processes (Parthasarathy et al., 1992). In this line, macrophages incubated with ox-LDL in pomegranate treated mice resulted in lowered superoxide anions release and an increase in the intrinsic antioxidant enzyme glutathione content than the control mice. Pomegranate also decreased accumulation of macrophage cholesterol and formation of foam cell by reducing macrophage binding, cell association and degradation of both native and ox-LDL most probably by the interaction of pomegranate flavonoids with LDL receptor (Aviram et al., 2000; Williams et al., 2004). Atherosclerotic lesions in pomegranate treated mice were smaller with fewer lipid-laden macrophage foam cells compared to control mice lesions (Aviram et al., 2000).

Polyphenolic rich pomegranate fruit reduces oxidative stress due to its good antioxidant scavenging abilities (Rummun et al., 2013). As such, Aviram et al. (2000) showed that the peel and mesocarp were more powerful in antioxidant abilities than the juice due to lower concentrations required to inhibit 50% LDL oxidation. Catechins and quercetin present in pomegranate juice have been shown to decrease atherosclerotic lesions, cellular uptake of LDL, LDL oxidation/aggregation, and to increase PON activity (Hayek et al., 1997). *In vitro* studies showed additional anti-atherogenic capacities of pomegranate juice by inhibiting LDL retention and aggregation, therefore preventing oxidation of LDL and foam cell formation as a consequence of uptake of aggregated LDL at an enhanced rate by macrophages (Aviram et al., 2000). LDL aggregation is inhibited due to binding of polyphenols to LDL via formation of ether bond (Hayek et al., 1997).

Anti-atherosclerotic properties of pomegranate juice were further substantiated by clinical studies in patients with carotid artery stenosis (CAS). Scavenging of reactive oxygen and nitrogen species (ROS/RNS) - that contribute to endothelium contraction- by pomegranate restored endothelial function leading to reduced systolic pressure and associated carotid intima media thickness (IMT) in CAS patients. Pomegranate juice also revealed its antioxidant potentials by decreasing serum oxidation status and increasing total antioxidant status, PON1 activity and glutathione content. Analysis of carotid intima from patients who underwent carotid endarterectomy showed reduced oxidative stress in subjects who consumed pomegranate juice compared to control, due to increased PON1 activity. The lesions were considered less atherogenic due smaller carotid lesion size; lower cholesterol and ox-LDL contents (Aviram et al., 2004). As such, pomegranate juice supplementation in obese volunteers showed that while fat mass in placebo group significantly increased compared to baseline, pomegranate juice halted the increase in individual weight and adiposity level (González-Ortiz et al., 2011). Furthermore, supplementation of 240 mL of pomegranate juice for 3 months has been reported to ameliorate stress-induced myocardial ischemia in patients with coronary heart disease as evidenced by a lowered myocardial ischemia and enhanced myocardial perfusion (Sumner et al., 2005). Daily supplementation of concentrated pomegranate polyphenol extract, over thirty days, significantly lowered plasma biomarkers for both lipid peroxidation and inflammation including IL-6 and hs-CRP, in obese people (Hosseini et al., 2016). One month daily intake of 240 ml/day of natural pomegranate juice showed a significant decrease in IL-6 level and soluble intercellular adhesion molecule-1, in adolescent patients with metabolic syndrome (Hashemi et al., 2010).

Diabetes

Under hyperglycemic conditions, which is usually the case in diabetic patients, excess of sugar molecules induce non-enzymatic glycation of proteins leading to formation of advanced glycation end products (AGEs) (Daroux et al., 2010). AGEs contribute to diabetic pathology as when they bound to their receptors, this interaction leads to over-production of ROS resulting to oxidative stress (Schaffer et al., 2012). Despite being rich in sugar, pomegranate did not alter LDL, HDL, triglycerides, fasting serum glucose or HbA1c levels in diabetic patients (Rock et al., 2008; Basu et al., 2013); most probably due to presence of sugar-containing polyphenolic tannins and anthocyanins such as punicalagin and cynanidin-3-glucoside respectively which have good antioxidant properties (Gil et al., 2000). However,

Esmaillzadeh et al. (2004) reported that consumption of pomegranate juice for a longer period exerted a hypocholesterolemic effect in diabetic patients with hyperlipidemia by decreasing total cholesterol and LDL levels, therefore increasing the HDL/LDL ratio following supplementation for 8 weeks (Esmaillzadeh et al., 2004). In addition to total cholesterol and LDL, supplementation of 200 mL of pomegranate juice reduced fasting blood sugar and MDA levels compared to baseline (Parsaeyan et al., 2012). Catechins have been reported to reduce solubility and thus absorption of cholesterol in rat intestine (Ikeda et al., 1992) while some isoflavones such as genistein and daidzein showed lowered expression and/or activities of enzymes involved in biosynthesis of LDL (Borradaile et al., 2002) and flavonoids such as naringenin and hesperetin increased cellular uptake and degradation of LDL by increasing expression of LDL receptor in HepG2 cells (Wilcox et al., 2001). Polyphenols from cocoa powder also decreased cholesterol level by increasing fecal excretion of cholesterol in high cholesterol-fed rat (Baba et al., 2007).

Consumption of pomegranate juice by diabetic patients attenuated oxidative stress state by counteracting the increased lipid peroxides and TBARS; and the decreased thiol groups. In addition to enhancing HDL-associated PON1 enzymatic activities as well as stabilizing PON1; pomegranate also increased serum FRAP concentration, reduced basal TBARS level and AAPH-induced serum lipid peroxidation (Rosenblat et al., 2006; Rock et al., 2008; Parsaeyan et al., 2012). *In vitro* activity of PON1 was increased in dose dependent manner in presence of increasing concentration of pomegranate juice suggesting that pomegranate juice component(s) act directly on PON1 (Rosenblat et al., 2006). Moreover, in hyperglycemic condition, HDL is prone to glycation resulting in decreased HDL-PON activity due to lipid and apoprotein compositional changes in HDL (Ferretti et al., 2001). The anti-glycation properties of pomegranate fruit, attributed to some polyphenols such as gallic and ellagic acids (Kumagai et al., 2015) may account for the inhibition of HDL glycation leading to PON activity. PON1 activity is also inactivated by the increased ox-LDL level during oxidative stress, most probably by interaction of oxidized lipid in ox-LDL with free thiol groups of PON1. Pomegranate polyphenols such as quercetin inhibited copper ion-induced LDL oxidation and associated PON1 inactivation (Aviram et al., 1999). Furthermore, quercetin was reported to increase PON1 mRNA expression and activity in hepatoma cell line (Gouédard et al., 2004). Pomegranate juice's component(s) may also interact with different scavenger receptors of ox-LDL, including CD36 due to its overexpression in response to ox-LDL cellular uptake (Hayek et al., 2005)- therefore counteracting the increase in uptake of ox-LDL by diabetic patients' human monocytes-derived macrophages (HMDM) compared to healthy person (Rosenblat et al., 2006). Pomegranate juice further decreased HMDM cellular lipid peroxides while increasing glutathione levels (Rosenblat et al., 2006). Consumption of pomegranate capsules exerted protective effects against ROS induced lipid peroxidation and counteracted the higher levels of plasma MDA and HNE in diabetic patient compared to healthy control (Basu et al., 2013).

Banihani et al., (2014) reported a significant decrease in fasting serum glucose level along with reduced insulin resistance. An increase in β-cell function was also observed three hours post pomegranate juice consumption. Furthermore, pomegranate decreased fasting plasma insulin level in addition to improving insulin resistance in volunteers at high cardiovascular disease risk following 4 weeks of pomegranate juice consumption (Tsang et al., 2012). Pomegranate phenolics can also cause insulin-like effects in glucose utilization. As such, pomegranate flower has been suggested to improve postprandial hyperglycemia in Zucker diabetic fatty rats by suppressing activity of α-glucosidase, thus inhibiting formation of monosaccharides from oligosaccharides and disaccharides leading to decreased gastrointestinal absorption of glucose (Li et al., 2005). Punicic acid, a compound present in pomegranate has been shown to activate peroxisome proliferator-activated receptor γ (PPAR γ) resulting in repression of TNF-α expression via antagonizing activities of NFκB in white adipose tissue and liver of db/db mice. Inhibition of TNF-α promotes insulin sensitivity thus decreases plasma glucose level (Hontecillas et al., 2009). Similarly, pomegranate seed lowered blood glucose level in streptozotocin diabetic rats (Das et al., 2001). In line with these studies, Shishehbor et al., (2016) further investigated the effect of pomegranate juice intake on subclinical inflammation type two diabetic patients and reported a significant decrease in patients' serum IL-6 level following four weeks juice consumption. However, contrary to the findings of Hosseini et al., (2016), and compared to obese patients, pomegranate juice did not modulate plasma TNF-α or hs-CRP levels in type two diabetic patients (Shishehbor et al., 2016). Pomegranate polyphenol may therefore selectively target and correct the balance of inflammatory mediators to physiological level, depending on the stage and progression of the inflammatory condition.

Hemodialysis

In hemodialytic patients, increased systemic inflammation and oxidative stress account for the pathogenesis of atherosclerosis and dysregulation of innate immunity. Peripheral polymorphonuclear leukocytes (PMNL) and monocytes release ROS and pro-inflammatory cytokines due to interaction of recurrent blood with the dialyzers. Pomegranate juice strengthened the innate immunity by decreasing the incidence of infections as a consequence of reduction in oxidative stress and inflammation levels. These reductions were associated with lowered PMNL priming as evidenced by the decrease in CD11b. PMNL degranulation produces myeloperoxidase (MPO) which is involved in production of hypochlorous acid (Shema-Didi et al., 2012). This highly reactive ROS oxidize proteins leading to the formation of advanced oxidation protein products (AOPP) (Witko-Sarsat et al., 2003). Patients who consumed pomegranate juice 1 h prior to their dialysis treatment three times per week for 1 year showed decrease levels of oxidative stress in terms of MPO, AOPP and MDA contents; and inflammation (IL-6 and TNF-α) compared to the placebo control group. Following a three month wash out period post one-year intervention, the level of serum inflammatory mediators returned back to the baseline level prior to the start of the intervention. Pomegranate juice also improved cardiovascular morbidity as evidenced by increased serum albumin; decreased AOPP and fibrinogen levels; and improved IMT in addition to structure and number of plaques of carotid arteries (Shema-Didi et al., 2012). The iron administered intravenously (IV) to treat anemia in patients during hemodialysis procedure also causes PMNL priming. Consumption of 100 mL pomegranate juice during the first hour of dialysis session reduced dialysis membrane- and iron-induced PMNL degranulation and disintegration which in turn reduced the inflammatory response. Analysis of blood biomarkers following dialysis session showed inhibition in IV iron-induced oxidative stress (decreased MPO and AOPP levels) aggravation. The polyphenolic richness of pomegranate juice may account for this protective effect due to its high antioxidant potentials including iron (II) chelating abilities (Shema-Didi et al., 2013; Rummun et al., 2013). Pomegranate also demonstrated its anti-atherogenic and anti-hypertensive properties in hemodialysis patients by decreasing systolic blood pressure, pulse pressure and triglyceride level; and increasing HDL level (Shema-Didi et al., 2014).

Hypertension

Supplementation of 50 mL of pomegranate juice for 2 weeks in hypertensive patients showed decreased serum angiotensin converting enzyme (ACE) activity by 36% and a minimal reduction of 5% in systolic blood pressure (Aviram and Dornfeld, 2001). The radical scavenging potential of pomegranate juice is accountable for the decrease in ACE activity as Usui et al. (1999) have shown that ACE activity is increased during oxidative stress. In the presence of ACE, angiotensin I is converted to angiotensin II which is involved in the promotion of atherosclerotic plaque at concentrations that do not affect blood pressure (Wojakowski et al., 2000). Following four weeks consumption of 500 mL pomegranate juice by volunteers at high cardiovascular disease risk resulted in a reduction in both systolic and diastolic blood pressures by inhibiting activity of 11β-hydroxysteroid dehydrogenase (11β-HSD); a reductase which converts cortisone to the active steroid cortisol. The latter has been reported to be associated with blood pressure (Tsang et al., 2012).

Cancer

Pomegranate juice supplementation by patients with recurrent prostate cancer led to a cytostatic effect as a consequence of slowed and not declined prostate-specific antigen (PSA) progression leading to prolongation of PSA doubling time (PDADT) and of disease stabilization. Pomegranate protective effect was not primarily hormonal as levels of androstenedione, dehydroepiandrosterone, estradiol, insulin-like growth factor, sex hormone-binding globulin and testosterone were unchanged. DNA is one of the important targets of ROS essential in tumour biology. ROS causes activation of transcription factors such as NFκB which induces expression of proto-oncogenes; genomic instability as a consequence of strand breakage, base modification and DNA oxidation; and protein oxidation, therefore enhancing the action of proteases which facilitates tumour invasion and metastasis (Toyokuni et al., 1995). Assessment of viability of LNCaP human prostate adenocarcinoma cells using patient serum instead of fetal bovine serum in the culture medium showed that serum of patients who consumed pomegranate juice reduced the growth of LNCaP (Pantuck et al., 2006). This suggests a chemopreventive role of pomegranate in cancer. Several flavonoids present in pomegranate such as luteolin, genistein, naringenin and quercetin may modulate estrogen-dependent cancers due to their estrogenic potencies (Zand et al., 2000; Kim et al., 2002).

Consumption of 200 mL of pomegranate juice by prostate cancer (PCa) or benign prostatic hyperplasia (BPH) patients for three days before prostatectomy, transurethral resection or adrenomectomy of prostate showed presence of ellagitannin derivatives (urolithin A glucuronide) in the prostate tissues at very low levels due to 18-24 h fasting before surgery resulting in excretion for several hours following an active enterohepatic circulation of the metabolites for short period of time. Urolithin A metabolites but not ellagitannins were also detected in plasma and urine. A poor expression of *p21* and associated rise in *c-Myc* was observed in PCa compared to BPH (González-Sarrías et al., 2010). p21 promotes G_2M arrest and apoptosis while the oncogene c-Myc represses the latter to induce cell growth and tumorigenesis (Albrecht et al., 2004; González-Sarrías et al., 2010). However, consumption of pomegranate did not alter expression of these genes as described *in vitro* (Albrecht et al., 2004) due to lower concentrations of polyphenolic metabolites reaching the target organ compared to the long period of treatment with high concentrations of whole pomegranate extracts *in vitro* (González-Sarrías et al., 2010). Ellagic acid, urolithin A and derivatives nonetheless have potentials in inhibiting the growth of human prostate cancer cells (Seeram et al., 2007) but the IC_{50} values were much higher than the concentration available in the prostate tissue *in vivo*. The product of DNA oxidation, 8-OHdg, levels in men with prostate cancer or benign tissues who consumed pomegranate powder in form of capsule for four weeks before radical prostatectomy were lowered compared to placebo group, with levels of urolithin A being negatively correlated with this DNA oxidation product (Freedland et al., 2013).

Rheumatoid Arthritis

Ghavipour and co-workers (2016) investigated the potential of pomegranate extract to alleviate disease condition in rheumatoid arthritis patients as part of a double blind randomized clinical trial. Supplementation with 40% ellagic acid containing pomegranate capsule over an eight week period was associated with a reduction in swollen and tender joints counts along with pain intensity in patients (Ghavipour et al., 2016). Furthermore, a six week supplementation with pomegranate juice in patients suffering with knee osteoarthritis showed that pomegranate juice consumption halted the disease progression by lowering serum matrix metallopeptidase (MMP)-13 level and a side by side amelioration in the patients antioxidant status; as evidenced by the elevated serum glutathione peroxidase level (Ghoochani et al., 2016).

Aging

Aging increases oxidatively damaged biomolecules as a consequence of increase ROS turnover (Bokov et al., 2004). A study conducted on elderly subjects showed that consumption of 250 mL of pomegranate juice (made from the pulp) for 4 weeks led to an increase in plasma FRAP, glutathione and catalase levels; and a decrease in plasma MDA and protein carbonyl contents. Since no alteration in ascorbic acid, vitamin E and reduced glutathione contents were observed, it was speculated that the reduction in oxidative stress state was attributed to the polyphenolic contents of pomegranate pulp. As such, quercetin, kaempferol and myricetin, which have good antioxidant capacities, were identified in the juice (Guo et al., 2008; Vinson et al., 1995; Hatia et al., 2014). Moreover, consumption of fried onions showed to increase basal quercetin level in plasma and associated total antioxidant capacity (McAnlis et al., 1999). Likewise, 19 of 25 glycosylated/acylated anthocyanins were detected in plasma of subjects following consumption of high-fat meal with blueberries that were liable to increase serum antioxidant status (Mazza et al., 2002).

BIOAVAILABILITY OF POMEGRANATE POLYPHENOLS

The antioxidant potentials of pomegranate juice have been attributed to its high polyphenolic content, 70% of which are ellagitannins with punicalagin being the most potent compound *in vitro* (Gil et al., 2000; Seeram et al., 2006). Consequently, punicalagin has been commonly used as an arbitrary reference to assess the protective effects of pomegranate against oxidative stress and inflammation *in vitro* (Adams et al., 2006; Chen et al., 2012). However, following oral consumption, the pomegranate bioactive compounds travel through different environmental conditions in various compartments of the digestive tract, starting with salivary and digestive enzymes, pH changes from the stomach to the small intestine and the gut microflora. Consequently, the chemical structure and conformation of the compounds are continuously changing, before being absorbed into the hepatic portal circulation (Figure 3). Adding to this, the absorbed compounds further undergo structural changes during the enterohepatic metabolism (Figure 4). It is now a well-established fact that the bioavailable bioactive compounds, following pomegranate polyphenol intake, exerts the therapeutic effect at the target site and have a completely different chemical scaffold from that of the ingested parent molecule. *In vitro* experiments showed that punicalagin is converted to punicalin and ellagic acid. However, *in vivo*, there is inefficient hydrolysis of punicalagin to ellagic acid as punicalin is further hydrolysed to gallagic acid which cyclized immediately forming gallagic acid dilactone. This compound is highly insoluble and was suggested to be inaccessible for hydrolysis by the gut microbiota. This was evidenced by the favoured formations of ellagic acid metabolites in subjects who consumed pomegranate with higher free ellagic acid compared to subjects consuming punicalagin richer pomegranate (Nuñez-Sánchez et al., 2014).

Ingestion of pomegranate juice concentrate by healthy human subjects showed that ellagitannins hydrolysed to ellagic acid in plasma was further bio-transformed by intestinal microflora to urolithin A and urolithin B within few hours (Figures 3, 4). Urolithins were further conjugated with glucuronyl, methyl and sulphate groups during phase II metabolism following first pass enterohepatic circulation, which are detected in both plasma and urine. Ellagic acids from hydrolysed ellagitannins also undergo methylation and glucuronosylation leading to formation of dimethylellagic acid glucuronide (Seeram et al., 2006). Metabolites of ellagitannins have been shown to accumulate in organs. For instance, Freedland et al. (2008) have reported that urolithin A, from oral administered pomegranate powder in form of capsule, accumulated in the prostate. Among the urolithin A metabolites, urolithin A sulfotransferases and methylated urolithin A were abundant in the prostate as well while urolithin A glucuronide was present in liver and kidney (Seeram et al., 2007). Urolithin A, urolithin B and their glucuronyl and sulphate conjugates (a total of 23 metabolites) were also reported to accumulate in human colon tissues in addition to being present in plasma and urine (Nuñez-Sánchez et al., 2014).

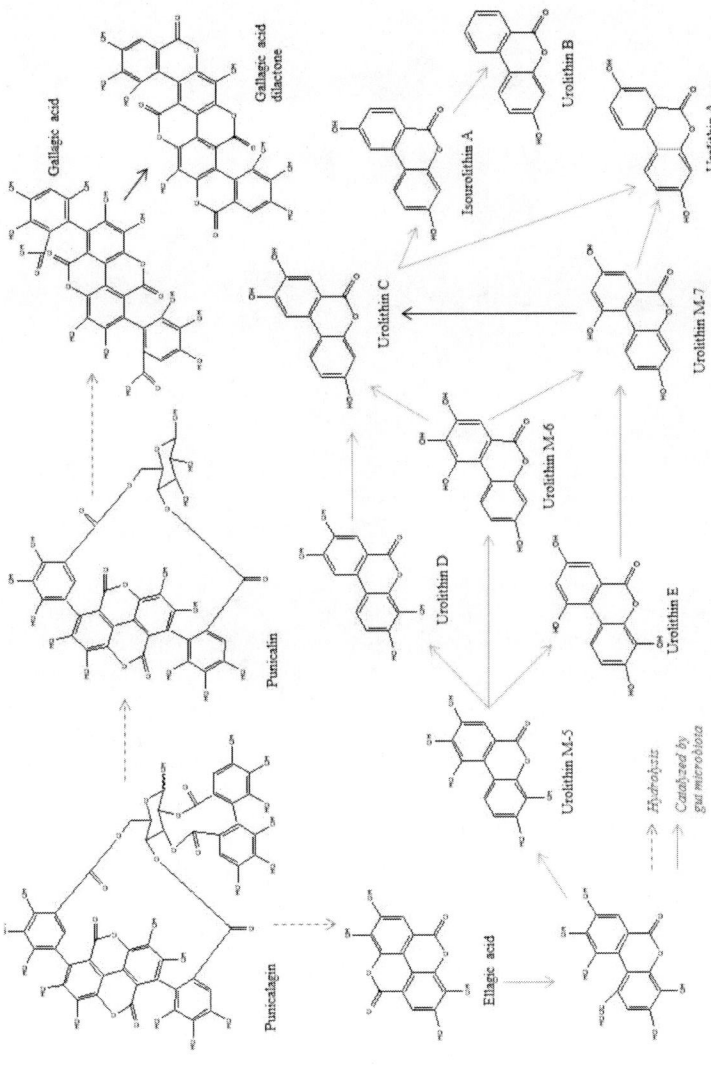

Figure 3. Urolithin metabolites formation from punicalagin (ellagitannins) by human gut microbiota (adapted from Nuñez-Sánchez et al., 2014).

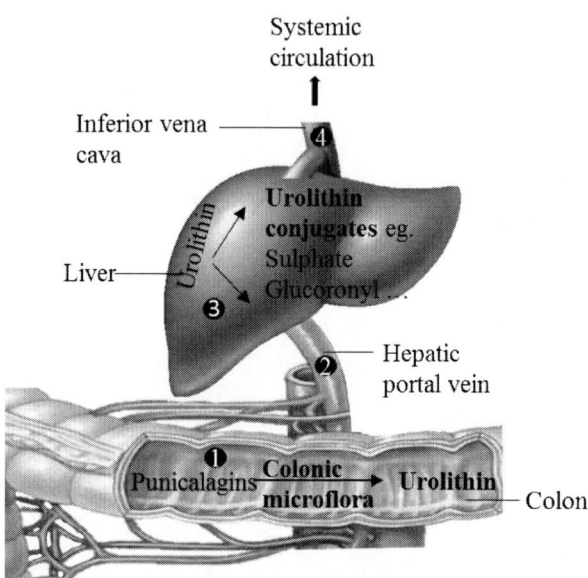

Figure 4. Pomegranate polyphenols bio-transformation and distribution via systemic circulation. ❶ Following oral ingestion pomegranate ellagitannins (e.g. punicalagins undergoes colonic bio-transformation to urolithin A, B, C and D under the action of colonic microbiota. ❷ The colonic metabolite enters the enterohepatic circulation via the hepatic portal vein and reaches the liver ❸ where they are further metabolized by the hepatic enzymes. ❹ The phase II conjugates enter the systemic circulation via the inferior vena cava and are distributed to the target organ to exert therapeutic effect or to get excreted in the urine.

An increasing number of studies emphasizing the beneficial impact of pomegranate on human health attributes this effect to pomegranate ellagitannin colonic metabolites rather than the parent phyto-constituents itself. Seeram et al., (2007) reported urolithins to exert greater growth inhibitory activity against both androgen-dependant and androgen-independent prostate cancer cells, as compared to pomegranate ellagic acid. Adding to this, urolithin A was reported to be a more potent inhibitor of myeloperoxidase and lactoperoxidase in both dose and time dependant manner compared to the parent ellagic acid (Saha et al., 2016). Furthermore, compared to ellagic acid treated mice, urolithin A treated mice showed lowered phorbol myristate acetate induced ear edema by a more significant volume, indicating that the anti-inflammatory effect of elagitannins rich extracts may be mediated via their colonic metabolites (Saha et al., 2016). Using in silico computational studies to predict the blood-brain barrier permeability of 21 pomegranate extract constituents (mostly ellagitannins), Yuan and co-workers (2016) revealed that none of the parent pomegranate constituents, except the colonic metabolite urolithin, fulfilled the required blood-brain barrier penetration criteria and therefore, hypothesized that urolithin may be the bioactive metabolite responsible for the reported efficacy of pomegranate extract against Alzheimer`s disease. Attempting to elucidate the active metabolite of ellagitannins which exerts anti-inflammatory effects, Giménez-Bastida et al., (2012) reported that only urolithin A-glucoronide (at ~ 15μM) significantly inhibited adhesion of monocytes to TNF-α stimulated human aortic endothelial cells compared to urolithin A, urolithin B-gluroconide and urolithin B, all of which had no effect on THP-1 monocytes adhesion. Furthermore, the inhibition potency of human aortic endothelial cells migration was shown to be greater for urolithin A-glucoronide and urolithin A compared to urolithin B-glucoronide, while urolithin B showed no significant inhibitory effect.

CONCLUSION

Pomegranate has been valued since ages and is still being used as ethnomedicine across the world. The polyphenolic richness; encompassing ellagitannins, phenolics, flavonoids and anthocyanins is accountable to the prophylactic effects of pomegranate showing antioxidant and anti-inflammatory potentials both *in vitro* and *in vivo*. Consumption of pomegranate juice by both healthy and diseased subjects has shown to attenuate the inflammatory state and to increase antioxidant status. As such, pomegranate polyphenols provide protective effects against the underlying pathophysiological hallmarks of a number of chronic diseases such as atherosclerosis, diabetes, hypertension, rheumatoid arthritis and cancer. In addition, the non-edible parts can be envisaged in supplements and food additives as a pragmatic approach to the prevention of some chronic diseases. However, the growing body of literature claims that bioactivity of the metabolised pomegranate polyphenols such as urolithins A, B and their conjugates rather than punicalagins which show promising therapeutic *in vitro*. The poor bioavailability due to the large size, slow lipid solubility, warrants further investigations of how to optimise the efficacy of punicalagins. The scientific evidence has validated the ethnomedicinal uses of the dietary nature of pomegranate and has helped to promote its use as a functional food to mitigate chronic human diseases.

REFERENCES

Abdel-Hady, N. M. (2013). Quantitative Diversity of Phenolic Content in Peels of Some Selected Egyptian Pomegranate Cultivars Correlated to Antioxidant and Anticancer Effects. *Journal of Applied Sciences Research*, *9*(8), 4823–4830.

Adams, L. S., Seeram, N. P., Aggarwal, B. B., Takada, Y., Sand, D., & Heber, D. (2006). Pomegranate Juice, Total Pomegranate Ellagitannins, and Punicalagin Suppress Inflammatory Cell Signaling in Colon Cancer Cells. *Journal of Agricultural and Food Chemistry*, *54*(3), 980–985.

Afaq, F., Saleem, M., Krueger, C. G., Reed, J. D., & Mukhtar, H. (2005). Anthocyanin- and hydrolyzable tannin-rich pomegranate fruit extract modulates MAPK and NF-κB pathways and inhibits skin tumorigenesis in CD-1 mice. *International Journal of Cancer*, *113*(3), 423–433.

Albrecht, M., Jiang, W., Kumi-Diaka, J., Lansky, E. P., Gommersall, L. M., Patel, A., Mansel, R. E., Neeman, I., Geldof, A. A., & Campbell, M. J. (2004). Pomegranate Extracts Potently Suppress Proliferation, Xenograft Growth, and Invasion of Human Prostate Cancer Cells. *Journal of Medicinal Food*, *7*(3), 274–283.

Ali, S. I., El-baz, F. K., El-emary, G. A. E., Khan, E. A., & Mohamed, A. A. (2014). HPLC-Analysis of Polyphenolic Compounds and Free Radical Scavenging Activity of Pomegranate Fruit (Punica granatum L.). *International Journal of Pharmaceutical and Clinical Research*, *6*(4), 348–355.

Aviram, M., Dornfeld, L., Rosenblat, M., Volkova, N., Kaplan, M., Coleman, R., Hayek, T., Presser, D., Fuhrman, B., 2000. Pomegranate juice consumption reduces oxidative stress, atherogenic modifications to LDL, and platelet aggregation: studies in humans and in atherosclerotic apolipoprotein E-deficient mice. *Am. J. Clin. Nutr. 71,* 1062–76.

Aviram, M., & Dornfeld, L. (2001). Pomegranate juice consumption inhibits serum angiotensin converting enzyme activity and reduces systolic blood pressure. *Atherosclerosis*, *158*(1), 195–198.

Aviram, M., Rosenblat, M., Billecke, S., Erogul, J., Sorenson, R., Bisgaier, C. L., Newton, R. S., & La Du, B. (1999). Human serum paraoxonase (PON 1) is inactivated by oxidized low density lipoprotein and preserved by antioxidants. *Free Radical Biology and Medicine*, *26*(7–8), 892–904.

Aviram, M., Rosenblat, M., Gaitini, D., Nitecki, S., Hoffman, A., Dornfeld, Volkova, N., Presser, D., Attias, J., Liker, H., & L., Hayek, T. (2004). Pomegranate juice consumption for 3 years by patients with carotid artery stenosis reduces common carotid intima-media thickness, blood pressure and LDL oxidation. *Clinical Nutrition, 23*(3), 423–433.

Baba, S., Natsume, M., Yasuda, A., Nakamura, Y., Tamura, T., Osakabe, N., Kanegae, M., & Kondo, K. (2007). Plasma LDL and HDL cholesterol and oxidized LDL concentrations are altered in normo- and hypercholesterolemic humans after intake of different levels of cocoa powder. *The Journal of Nutrition, 137*(6), 1436–1441.

Bagri, P., Ali, M., Aeri, V., Bhowmik, M., & Sultana, S. (2009). Antidiabetic effect of Punica granatum flowers: effect on hyperlipidemia, pancreatic cells lipid peroxidation and antioxidant enzymes in experimental diabetes. *Food and Chemical Toxicology: An International Journal Published for the British Industrial Biological Research Association, 47*(1), 50–4.

Balasundram, N., Sundram, K., & Samman, S. (2006). Phenolic compounds in plants and agri-industrial by-products: Antioxidant activity, occurrence, and potential uses. *Food Chemistry, 99*(1), 191–203.

Banihani, S. A., Makahleh, S. M., El-Akawi, Z., Al-Fashtaki, R. A., Khabour, O. F., Gharibeh, M. Y., Saadah, N. A., Al-Hashimi, F. H., & Al-Khasieb, N. J. (2014). Fresh pomegranate juice ameliorates insulin resistance, enhances β-cell function, and decreases fasting serum glucose in type 2 diabetic patients. *Nutrition Research, 34*(10), 862–867.

Baraliakos, X., Listing, J., Fritz, C., Haibel, H., Alten, R., Burmester, G. R., Krause, A., Schewe, S., Schneider, M., Sorensen, H., Schmidt, R., Sieper, J., & Braun, J. (2011). Persistent clinical efficacy and safety of infliximab in ankylosing spondylitis after 8 years-early clinical response predicts long-term outcome. *Rheumatology, 50*(9), 1690–1699.

Basu, A., Newman, E. D., Bryant, A. L., Lyons, T. J., & Betts, N. M. (2013). Pomegranate Polyphenols Lower Lipid Peroxidation in Adults with Type 2 Diabetes but Have No Effects in Healthy Volunteers: A Pilot Study. *Journal of Nutrition and Metabolism, 2013*, 1–7.

Bensaad, L., & Kim, K. (2015). Phytochemical Constituents and Analgesic Activity of Ethyl Acetate Fraction of Punicagranatum L (Punicaceae). *Tropical Journal of Pharmaceutical Research, 14*(1), 87-93.

Biswas, S. K. (2016). Does the Interdependence between Oxidative Stress and Inflammation Explain the Antioxidant Paradox? *Oxidative Medicine and Cellular Longevity, 2016*, 1–9.

Bokov, A., Chaudhuri, A., & Richardson, A. (2004). The role of oxidative damage and stress in aging. *Mechanisms of Ageing and Development*, *125*(10–11), 811–826.

Bonomini, F., Tengattini, S., Fabiano, A., Bianchi, R., & Rezzani, R. (2008). Atherosclerosis and oxidative stress. *Histology and Histopathology*, *23*(3), 381–90.

Bonzanini, F., Bruni, R., Palla, G., Serlataite, N., & Caligiani, A. (2009). Identification and distribution of lignans in Punica granatum L. fruit endocarp, pulp, seeds, wood knots and commercial juices by GC–MS. *Food Chemistry*, *117*(4), 745–749.

Borradaile, N. M., Dreu, L. E. De, Wilcox, L. J., Edwards, J. Y., & Huff, M. W. (2002). Soya phytoestrogens, genistein and daidzein, decrease apolipoprotein B secretion from HepG2 cells through multiple mechanisms. *Biochemical Journal*, *366*(2), 531–539.

Çam, M., Hışıl, Y., & Durmaz, G. (2009). Classification of eight pomegranate juices based on antioxidant capacity measured by four methods. *Food Chemistry*, *112*(3), 721–726.

Chames, P., Van Regenmortel, M., Weiss, E., & Baty, D. (2009). Therapeutic antibodies: successes, limitations and hopes for the future. *British Journal of Pharmacology*, *157*(2), 220–233.

Chen, B., Tuuli, M. G., Longtine, M. S., Shin, J. S., Lawrence, R., Inder, T., & Michael Nelson, D. (2012). Pomegranate juice and punicalagin attenuate oxidative stress and apoptosis in human placenta and in human placental trophoblasts. *AJP: Endocrinology and Metabolism*, *302*(9), E1142–E1152.

Cooke, M. S., Evans, M. D., Dizdaroglu, M., & Lunec, J. (2003). Oxidative DNA damage: mechanisms, mutation, and disease. *The FASEB Journal*, *17*(10), 1195–1214.

Crascì, L., Lauro, M. R., Puglisi, G., & Panico, A. (2016). Natural Antioxidant Polyphenols On Inflammation Management: Anti-glycation Activity Vs Metalloproteinases Inhibition. *Critical Reviews in Food Science and Nutrition*, *56*, Doi.org/10.1080/10408398.2016.1229657.

Cuvelier, M. E., Richard, H., & Berset, C. (1992). Comparison of t he Antioxidative Activity of Some Acid-phenols: Structure-Activity Relationship. *Bioscience, Biotechnology, and Biochemistry*, *56*(2), 324–325.

Daglia, M. (2012). Polyphenols as antimicrobial agents. *Current Opinion in Biotechnology*, *23*(2), 174–181.

Dalle-Donne, I. (2006). Biomarkers of Oxidative Damage in Human Disease. *Clinical Chemistry*, *52*(4), 601–623.

Daroux, M., Prévost, G., Maillard-Lefebvre, H., Gaxatte, C., D'Agati, V. D., Schmidt, A. M., & Boulanger, É. (2010). Advanced glycation end-products: Implications for diabetic and non-diabetic nephropathies. *Diabetes & Metabolism*, *36*(1), 1–10.

Das, A. K., Mandal, S. C., Banerjee, S. K., Sinha, S., Saha, B. P., & Pal, M. (2001). Studies on the hypoglycaemic activity of Punica granatum seed in streptozotocin induced diabetic rats. *Phytotherapy Research*, *15*(7), 628–629.

Donath, M. Y., & Shoelson, S. E. (2011). Type 2 diabetes as an inflammatory disease. *Nature Reviews. Immunology*, *11*(2), 98–107.

Duke, J. A.; Ayensu, E. S. *Medicinal Plants of China*; Western Pacific Series No. 2; WHO Regional Publications, Manila, PHILIPPINES, 1985; pp 243.

Elalouf, O., & Elkayam, O. (2015). Long-term safety and efficacy of infliximab for the treatment of ankylosing spondylitis. *Therapeutics and Clinical Risk Management*, *11*, 1719-1726.

Esmaillzadeh, A., Tahbaz, F., Gaieni, I., Alavi-Majd, H., & Azadbakht, L. (2004). Concentrated Pomegranate Juice Improves Lipid Profiles in Diabetic Patients with Hyperlipidemia. *Journal of Medicinal Food*, *7*(3), 305–308.

Ferretti, G., Bacchetti, T., Marchionni, C., Caldarelli, L., & Curatola, G. (2001). Effect of glycation of high density lipoproteins on their physicochemical properties and on paraoxonase activity. *Acta Diabetologica*, *38*(4), 163–169.

Fischer, U. A., Carle, R., & Kammerer, D. R. (2011). Identification and quantification of phenolic compounds from pomegranate (Punica granatum L.) peel, mesocarp, aril and differently produced juices by HPLC-DAD–ESI/MSn. *Food Chemistry*, *127*(2), 807–821.

Freedland, S. J., Carducci, M., Kroeger, N., Partin, A., Rao, J. Y., Jin, Y., Kerkoutian, S., Wu, H., Li, Y., Creel, P., Mundy, K., Gurganus, R., Fedor, H., King, S. A., Zhang, Y., Heber, D., & Pantuck, A. J. (2013). A Double-Blind, Randomized, Neoadjuvant Study of the Tissue Effects of POMx Pills in Men with Prostate Cancer Before Radical Prostatectomy. *Cancer Prevention Research*, *6*(10), 1120–1127.

Freedman, J. E. (2008). Oxidative Stress and Platelets. *Arteriosclerosis, Thrombosis, and Vascular Biology*, *28*(3), s11–s16.

Fuhrman, B., Oiknine, J., & Aviram, M. (1994). Iron induces lipid peroxidation in cultured macrophages, increases their ability to oxidatively modify LDL, and affects their secretory properties. *Atherosclerosis*, *111*(1), 65–78.

Fullerton, J. N., & Gilroy, D. W. (2016). Resolution of inflammation: a new therapeutic frontier. *Nature Reviews. Drug Discovery*, *15*(8), 551–67.

Ghavipour, M., Sotoudeh, G., Tavakoli, E., Mowla, K., Hasanzadeh, J., & Mazloom, Z. (2016). Pomegranate extract alleviates disease activity and some blood biomarkers of inflammation and oxidative stress in Rheumatoid Arthritis patients. *European Journal of Clinical Nutrition*, (February), 1–5.

Ghoochani, N., Karandish, M., Mowla, K., Haghighizadeh, M. H., & Jalali, M. T. (2016). The effect of pomegranate juice on clinical signs, matrix metalloproteinases and antioxidant status in patients with knee osteoarthritis. *Journal of the Science of Food and Agriculture*, *96*(13), 4377–4381.

Gil, M. I., Tomás-Barberán, F. A., Hess-Pierce, B., Holcroft, D. M., & Kader, A. A. (2000). Antioxidant activity of pomegranate juice and its relationship with phenolic composition and processing. *Journal of Agricultural and Food Chemistry*, *48*(10), 4581–9.

Giménez-Bastida, J. A., González-Sarrías, A., Larrosa, M., Tomás-Barberán, F., Espín, J. C., & García-Conesa, M. T. (2012). Ellagitannin metabolites, urolithin A glucuronide and its aglycone urolithin A, ameliorate TNF-α-induced inflammation and associated molecular markers in human aortic endothelial cells. *Molecular Nutrition & Food Research*, *56*(5), 784–796.

Gloire, G., Legrand-Poels, S., & Piette, J. (2006). NF-κB activation by reactive oxygen species: Fifteen years later. *Biochemical Pharmacology*, *72*(11), 1493–1505.

González-Ortiz, M., Martínez-Abundis, E., Espinel-Bermúdez, M. C., & Pérez-Rubio, K. G. (2011). Effect of Pomegranate Juice on Insulin Secretion and Sensitivity in Patients with Obesity. *Annals of Nutrition and Metabolism*, *58*(3), 220–223.

González-Sarrías, A., Giménez-Bastida, J. A., García-Conesa, M. T., Gómez-Sánchez, M. B., García-Talavera, N. V., Gil-Izquierdo, A., Sánchez-Álvarez, C., Fontana-Compiano, L. O., Morga-Egea, J. P., Pastor-Quirante, F. A., Martínez-Díaz, F., Tomás-Barberán, F. A., & Espín, J. C. (2010). Occurrence of urolithins, gut microbiota ellagic acid metabolites and proliferation markers expression response in the human prostate gland upon consumption of walnuts and pomegranate juice. *Molecular Nutrition & Food Research*, *54*(3), 311–322.

González-Trujano, M. E., Pellicer, F., Mena, P., Moreno, D. A., & García-Viguera, C. (2015). Antinociceptive and anti-inflammatory activities of a pomegranate (Punica granatum L.) extract rich in ellagitannins. *International Journal of Food Sciences and Nutrition, 66*(4), 395–399.

González, R., Ballester, I., López-Posadas, R., Suárez, M. D., Zarzuelo, A., Martínez-Augustin, O., & Medina, F. S. De. (2011). Effects of Flavonoids and other Polyphenols on Inflammation. *Critical Reviews in Food Science and Nutrition, 51*(4), 331–362.

Gouédard, C., Barouki, R., & Morel, Y. (2004). Dietary Polyphenols Increase Paraoxonase 1 Gene Expression by an Aryl Hydrocarbon Receptor-Dependent Mechanism. *Molecular and Cellular Biology, 24*(12), 5209–5222.

Granado-Serrano, A. B., Martín, M. A., Bravo, L., Goya, L., & Ramos, S. (2012). Quercetin modulates Nrf2 and glutathione-related defenses in HepG2 cells: Involvement of p38. *Chemico-Biological Interactions, 195* (2), 154–164.

Guo, C., Wei, J., Yang, J., Xu, J., Pang, W., & Jiang, Y. (2008). Pomegranate juice is potentially better than apple juice in improving antioxidant function in elderly subjects. *Nutrition Research, 28*(2), 72–77.

Haraoui, B., & Bykerk, V. (2007). Etanercept in the treatment of rheumatoid arthritis. *Therapeutics and Clinical Risk Management, 3*(1), 99–105.

Hashemi, M., Kelishadi, R., Hashemipour, M., Zakerameli, A., Khavarian, N., Ghatrehsamani, S., & Poursafa, P. (2010). Acute and long-term effects of grape and pomegranate juice consumption on vascular reactivity in paediatric metabolic syndrome. *Cardiology in the Young, 20*(1), 73-77.

Hatia, S., Septembre-Malaterre, A., Le Sage, F., Badiou-Bénéteau, A., Baret, P., Payet, B., Lefebvre d'hellencourt, C., & Gonthier, M. P. (2014). Evaluation of antioxidant properties of major dietary polyphenols and their protective effect on 3T3-L1 preadipocytes and red blood cells exposed to oxidative stress. *Free Radical Research, 48*(4), 387–401.

Hayek, T., Fuhrman, B., Vaya, J., Rosenblat, M., Belinky, P., Coleman, R., Elis, A., Aviram, M., 1997. Reduced Progression of Atherosclerosis in Apolipoprotein E Deficient Mice Following Consumption of Red Wine, or Its Polyphenols Quercetin or Catechin, Is Associated With Reduced Susceptibility of LDL to Oxidation and Aggregation. *Arterioscler. Thromb. Vasc. Biol.* 17, 2744–2752.

Hayek, T., Hussein, K., Aviram, M., Coleman, R., Keidar, S., Pavoltzky, E., & Kaplan, M. (2005). Macrophage-foam cell formation in streptozotocin-induced diabetic mice: Stimulatory effect of glucose. *Atherosclerosis, 183*(1), 25–33.

Hontecillas, R., O'Shea, M., Einerhand, A., Diguardo, M., & Bassaganya-Riera, J. (2009). Activation of PPAR γ and α by Punicic Acid Ameliorates Glucose Tolerance and Suppresses Obesity-Related Inflammation. *Journal of the American College of Nutrition, 28*(2), 184–195.

Hosseini, B., Saedisomeolia, A., Wood, L.G., Yaseri, M., Tavasoli, S., 2016. Effects of pomegranate extract supplementation on inflammation in overweight and obese individuals: A randomized controlled clinical trial. *Complementary Therapies in Clinical Practice*. 22, 44–50.

Ikeda, I., Imasato, Y., Sasaki, E., Nakayama, M., Nagao, H., Takeo, T., Yayabe, F., & Sugano, M. (1992). Tea catechins decrease micellar solubility and intestinal absorption of cholesterol in rats. *Biochimica et Biophysica Acta (BBA) - Lipids and Lipid Metabolism, 1127*(2), 141–146.

Jaiswal, V., DerMarderosian, A., & Porter, J. R. (2010). Anthocyanins and polyphenol oxidase from dried arils of pomegranate (Punica granatum L.). *Food Chemistry, 118*(1), 11–16.

Joseph, S. V., Edirisinghe, I., & Burton-Freeman, B. M. (2016). Fruit Polyphenols: A Review of Anti-inflammatory Effects in Humans. *Critical Reviews in Food Science and Nutrition, 56*(3), 419–444.

Jurenka, J. S. (2008). Therapeutic applications of pomegranate (Punica granatum L.): a review. *Alternative Medicine Review: A Journal of Clinical Therapeutic, 13*(2), 128–44.

Kenne, E., Erlandsson, A., Lindbom, L., Hillered, L., & Clausen, F. (2012). Neutrophil depletion reduces edema formation and tissue loss following traumatic brain injury in mice. *Journal of Neuroinflammation, 9*(1), doi: 10.1186/1742-2094-9-17.

Kidd, B. L., & Urban, L. A. (2001). Mechanisms of inflammatory pain. *British Journal of Anaesthesia, 87*(1), 3–11.

Kim, N. D., Mehta, R., Yu, W., Neeman, I., Livney, T., Amichay, A., Poirier, D., Nicholls, P., Kirby, A., Jiang, W., Mansel, R., Ramachandran, C., Rabi, T., Kaplan, B., & Lansky, E. (2002). Chemopreventive and adjuvant therapeutic potential of pomegranate (Punica granatum) for human breast cancer. *Breast Cancer Research and Treatment, 71*(3), 203–17.

Kumagai, Y., Nakatani, S., Onodera, H., Nagatomo, A., Nishida, N., Matsuura, Y., Kobata, K., & Wada, M. (2015). Anti-Glycation Effects of Pomegranate (Punica granatum L.) Fruit Extract and Its Components in Vivo and in Vitro. *Journal of Agricultural and Food Chemistry*, *63*(35), 7760–7764.

Labib, R. M., & El-Ahmady, S. H. (2015). Antinociceptive, anti-gastric ulcerogenic and anti-inflammatory activities of standardized Egyptian pomegranate peel extract. *Journal of Applied Pharmaceutical Science*, *5*(1), 48–51.

Lago, J., Toledo-Arruda, A., Mernak, M., Barrosa, K., Martins, M., Tibério, I., & Prado, C. (2014). Structure-Activity Association of Flavonoids in Lung Diseases. *Molecules*, *19*(3), 3570–3595.

Lansky, E. P., & Newman, R. A. (2007). Punica granatum (pomegranate) and its potential for prevention and treatment of inflammation and cancer. *Journal of Ethnopharmacology*, *109*(2), 177–206.

Leporatti, M., & Ghedira, K. (2009). Comparative analysis of medicinal plants used in traditional medicine in Italy and Tunisia. *Journal of Ethnobiology and Ethnomedicine*, *5*(1), doi: 10.1186/1746-4269-5-31.

Li, Y., Wen, S., Kota, B. P., Peng, G., Li, G. Q., Yamahara, J., & Roufogalis, B. D. (2005). Punica granatum flower extract, a potent α-glucosidase inhibitor, improves postprandial hyperglycemia in Zucker diabetic fatty rats. *Journal of Ethnopharmacology*, *99*(2), 239–244.

Malešev, D., & Kuntić, V. (2007). Investigation of metal-flavonoid chelates and the determination of flavonoids via metal-flavonoid complexing reactions. *Journal of the Serbian Chemical Society*, *72*(10), 921–939.

Mantovani, A., Allavena, P., Sica, A., & Balkwill, F. (2008). Cancer-related inflammation. *Nature*, *454*(7203), 436–444.

Marques, L., Pinheiro, A., Araújo, J., de Oliveira, R., Silva, S., Abreu, I., de Sousa, E., Fernandes, E., Luchessi, A., Silbiger, V., Nicolete, R., & Lima-Neto, L. (2016). Anti-Inflammatory Effects of a Pomegranate Leaf Extract in LPS-Induced Peritonitis. *Planta Medica*, 3–7.

Maskrey, B. H., Megson, I. L., Whitfield, P. D., & Rossi, A. G. (2011). Mechanisms of resolution of inflammation: A focus on cardiovascular disease. *Arteriosclerosis, Thrombosis, and Vascular Biology*, *31*(5), 1001–1006.

Matthaiou, C. M., Goutzourelas, N., Stagos, D., Sarafoglou, E., Jamurtas, A., Koulocheri, S. D., Haroutounian, S. A., Tsatsakis, A. M., & Kouretas, D. (2014). Pomegranate juice consumption increases GSH levels and reduces lipid and protein oxidation in human blood. *Food and Chemical Toxicology*, *73*, 1–6.

Mazza, G., Kay, C. D., Cottrell, T., & Holub, B. J. (2002). Absorption of Anthocyanins from Blueberries and Serum Antioxidant Status in Human Subjects. *Journal of Agricultural and Food Chemistry*, *50*(26), 7731–7737.

McAnlis, G. T., McEneny, J., Pearce, J., & Young, I. S. (1999). Absorption and antioxidant effects of quercetin from onions, in man. *European Journal of Clinical Nutrition*, *53*(2), 92–96.

Mogarekar, M. R., Dhabe, M. G., & Gujrathi, C. C. (2016). A study of paraoxonase1 (PON1) activities, HDL cholesterol and its association with vascular complication in type 2 diabetes mellitus. *International Journal of Diabetes in Developing Countries*, *1*, doi:10.1007/s13410-016-0465-x.

Mohammad, S. M., Kashani, H. H., 2012. Chemical composition of the plant Punica granatum L. (Pomegranate) and its effect on heart and cancer. *Journal of Medicinal Plants Research*. 6, 5306–5310.

Moskaug, J. Ø., Carlsen, H., Myhrstad, M. C. W., & Blomhoff, R. (2005). Polyphenols and glutathione synthesis regulation. *The American Journal of Clinical Nutrition*, *81*(1 Suppl), 277S–283S.

Nuñez-Sánchez, M. A., García-Villalba, R., Monedero-Saiz, T., García-Talavera, N. V., Gómez-Sánchez, M. B., Sánchez-Álvarez, C., García-Albert, A. M., Rodríguez-Gil. F. J., Ruiz-Marín, M., Pastor-Quirante, F. A., Martínez-Díaz, F., Yáñez-Gascón, M. j., González-Sarrías, A., Tomás-Barberán, F. A., & Espín, J. C. (2014). Targeted metabolic profiling of pomegranate polyphenols and urolithins in plasma, urine and colon tissues from colorectal cancer patients. *Molecular Nutrition & Food Research*, *58*(6), 1199–1211.

Pantuck, A. J., Leppert, J. T., Zomorodian, N., Aronson, W., Hong, J., Barnard, R. J., Seeram, N., Liker, H., Wang, H., Elashoff, R., Heber, D., Aviram, M., Ignarro, L., & Belldegrun, A. (2006). Phase II study of pomegranate juice for men with rising prostate-specific antigen following surgery or radiation for prostate cancer. *Clinical Cancer Research: An Official Journal of the American Association for Cancer Research*, *12*(13), 4018–26.

Parsaeyan, N., Mozaffari–Khosravi, H., & Mozayan, M. (2012). Effect of pomegranate juice on paraoxonase enzyme activity in patients with type 2 diabetes. *Journal of Diabetes & Metabolic Disorders*, *11*(1), doi: 10.1186/ 2251-6581-11-11.

Parthasarathy, S., Raghavamenon, A., Garelnabi, M.O., Santanam, N., 2010. Oxidized Low-Density Lipoprotein, in: Uppu, R.M., Murthy, S.N., Pryor, W.A., Parinandi, N.L. (Eds.), *Free Radicals and Antioxidant Protocols, Methods in Molecular Biology.* Humana Press, Totowa, NJ, pp. 403–417.

Patel, A., Patel, A., Patel, A., & Patel, N. (2010). Determination of polyphenols and free radical scavenging activity of Tephrosia purpurea linn leaves (Leguminosae). *Pharmacognosy Research*, *2*(3), 152-8.

Patel, C., Dadhaniya, P., Hingorani, L., & Soni, M. G. (2008). Safety assessment of pomegranate fruit extract: Acute and subchronic toxicity studies. *Food and Chemical Toxicology*, *46*(8), 2728–2735.

Procházková, D., Boušová, I., & Wilhelmová, N. (2011). Antioxidant and prooxidant properties of flavonoids. *Fitoterapia*, *82*(4), 513–523.

Puchner, A., & Blüml, S. (2015). IL-6 blockade in chronic inflammatory diseases. *Wiener Medizinische Wochenschrift*, *165*(1–2), 14–22.

Quan, L. D., Thiele, G. M., Tian, J., & Wang, D. (2008). The development of novel therapies for rheumatoid arthritis. *Expert Opinion on Therapeutic Patents*, *18*(7), 723–738.

Reuter, S., Gupta, S. C., Chaturvedi, M. M., & Aggarwal, B. B. (2010). Oxidative stress, inflammation, and cancer: How are they linked? *Free Radical Biology and Medicine*, *49*(11), 1603–1616.

Robinson, W. H., Lepus, C. M., Wang, Q., Raghu, H., Mao, R., Lindstrom, T. M., & Sokolove, J. (2016). Low-grade inflammation as a key mediator of the pathogenesis of osteoarthritis. *Nature Reviews Rheumatology*, *12*(10), 580–592.

Rock, W., Rosenblat, M., Miller-Lotan, R., Levy, A. P., Elias, M., & Aviram, M. (2008). Consumption of Wonderful Variety Pomegranate Juice and Extract by Diabetic Patients Increases Paraoxonase 1 Association with High-Density Lipoprotein and Stimulates Its Catalytic Activities. *Journal of Agricultural and Food Chemistry*, *56*(18), 8704–8713.

Rosenblat, M., Hayek, T., & Aviram, M. (2006). Anti-oxidative effects of pomegranate juice (PJ) consumption by diabetic patients on serum and on macrophages. *Atherosclerosis*, *187*(2), 363–371.

Rummun, N., Somanah, J., Ramsaha, S., Bahorun, T., & Neergheen-Bhujun, V. S. (2013). Bioactivity of Nonedible Parts of Punica granatum L.: A Potential Source of Functional Ingredients. *International Journal of Food Science*, *2013*, 1–12.

Saad, L. Ben, Hwi, K. K., & Quah, T. (2014). Evaluation of the antinociceptive effect of the ethanolic extract of Punica granatum. *African Journal of Traditional, Complementary, and Alternative Medicines: AJTCAM*, *11*(3), 228–33.

Saha, P., Yeoh, B. S., Singh, R., Chandrasekar, B., Vemula, P. K., Haribabu, B., Vijay-Kumar, M., & Jala, V. R. (2016). Gut Microbiota Conversion of Dietary Ellagic Acid into Bioactive Phytoceutical Urolithin A Inhibits Heme Peroxidases. *PLOS ONE*, *11*(6), doi: 10.1371/journal.pone. 0156811.

Salwe, K. J., & Sachdev, D. (2014). Evaluation of antinociceptive and anti-inflammatory effect of the hydroalcoholic extracts of leaves and fruit peel of P. Granatum in experimental animals. *Asian Journal of Pharmaceutical and Clinical Research*, *7*(SUPPL. 2), 137–141.

Sarker, M., Das, S. C., Saha, S. K., Al Mahmud, Z., & Bachar, S. C. (2012). Analgesic and anti-inflammatory activities of flower extracts of Punica granatum Linn. (Punicaceae). *Journal of Applied Pharmaceutical Science*, *2*(4), 133–136.

Sawant, D. A., Wilson, R. L., Tharakan, B., Stagg, H. W., Hunter, F. A., & Childs, E. W. (2014). Tumor necrosis factor-α-induced microvascular endothelial cell hyperpermeability: role of intrinsic apoptotic signaling. *Journal of Physiology and Biochemistry*, *70*(4), 971–80.

Schaffer, S. W., Jong, C. J., & Mozaffari, M. (2012). Role of oxidative stress in diabetes-mediated vascular dysfunction: Unifying hypothesis of diabetes revisited. *Vascular Pharmacology*, *57*(5–6), 139–149.

Schubert, S. Y., Lansky, E. P., & Neeman, I. (1999). Antioxidant and eicosanoid enzyme inhibition properties of pomegranate seed oil and fermented juice flavonoids. *Journal of Ethnopharmacology*, *66*(1), 11–17.

Seeram, N. P., Aronson, W. J., Zhang, Y., Henning, S. M., Moro, A., Lee, R., Sartippour, M., Harris, D. M., Rettig, M., Suchard, M. A., Pantuck, A. J., Belldegrun, A., & Heber, D. (2007). Pomegranate Ellagitannin-Derived Metabolites Inhibit Prostate Cancer Growth and Localize to the Mouse Prostate Gland. *Journal of Agricultural and Food Chemistry*, *55*(19), 7732–7737.

Seeram, N. P., Henning, S. M., Zhang, Y., Suchard, M., Li, Z., & Heber, D. (2006). Pomegranate juice ellagitannin metabolites are present in human plasma and some persist in urine for up to 48 hours. *The Journal of Nutrition, 136*(10), 2481–5.

Shema-Didi, L., Kristal, B., Ore, L., Shapiro, G., Geron, R., & Sela, S. (2013). Pomegranate juice intake attenuates the increase in oxidative stress induced by intravenous iron during hemodialysis. *Nutrition Research, 33*(6), 442–446.

Shema-Didi, L., Kristal, B., Sela, S., Geron, R., & Ore, L. (2014). Does Pomegranate intake attenuate cardiovascular risk factors in hemodialysis patients? *Nutrition Journal, 13*(1), doi: 10.1186/1475-2891-13-18.

Shema-Didi, L., Sela, S., Ore, L., Shapiro, G., Geron, R., Moshe, G., & Kristal, B. (2012). One year of pomegranate juice intake decreases oxidative stress, inflammation, and incidence of infections in hemodialysis patients: A randomized placebo-controlled trial. *Free Radical Biology and Medicine, 53*(2), 297–304.

Shishehbor, F., Mohammad shahi, M., Zarei, M., Saki, A., Zakerkish, M., Shirani, F., & Zare, M. (2016). Effects of Concentrated Pomegranate Juice on Subclinical Inflammation and Cardiometabolic Risk Factors for Type 2 Diabetes: A Quasi-Experimental Study. *International Journal of Endocrinology and Metabolism, 14*(1), doi: 10.5812/ijem.33835.

Solak, Y., Afsar, B., Vaziri, N. D., Aslan, G., Yalcin, C. E., Covic, A., & Kanbay, M. (2016). Hypertension as an autoimmune and inflammatory disease. *Hypertension Research, 39*(8), 567–573.

Sreekumar, S., Sithul, H., Muraleedharan, P., Azeez, J. M., & Sreeharshan, S. (2014). Pomegranate Fruit as a Rich Source of Biologically Active Compounds. *BioMed Research International, 2014*, 1–12.

Stover, E., & Mercure, E. W. (2007). The pomegranate: A new look at the fruit of paradise. *HortScience, 42*(5), 1088–1092.

Sumner, M. D., Elliott-Eller, M., Weidner, G., Daubenmier, J. J., Chew, M. H., Marlin, R., Raisin, C. J., & Ornish, D. (2005). Effects of Pomegranate Juice Consumption on Myocardial Perfusion in Patients With Coronary Heart Disease. *The American Journal of Cardiology, 96*(6), 810–814.

Toyokuni, S., Okamoto, K., Yodoi, J., & Hiai, H. (1995). Persistent oxidative stress in cancer. *FEBS Letters, 358*(1), 1–3.

Tsang, C., Smail, N. F., Almoosawi, S., Davidson, I., & Al-Dujaili, E. A. S. (2012). Intake of polyphenol-rich pomegranate pure juice influences urinary glucocorticoids, blood pressure and homeostasis model assessment of insulin resistance in human volunteers. *Journal of Nutritional Science*, *1*, doi: 10.1017/jns.2012.10.

Uchida, K. (2003). 4-Hydroxy-2-nonenal: a product and mediator of oxidative stress. *Progress in Lipid Research*, *42*(4), 318–343.

Usui, M., Egashira, K., Kitamoto, S., Koyanagi, M., Katoh, M., Kataoka, C., Shimokawa, H., & Takeshita, A. (1999). Pathogenic Role of Oxidative Stress in Vascular Angiotensin-Converting Enzyme Activation in Long-Term Blockade of Nitric Oxide Synthesis in Rats. *Hypertension*, *34*(4), 546–551.

Vinson, J. A., Dabbagh, Y. A., Serry, M. M., & Jang, J. (1995). Plant Flavonoids, Especially Tea Flavonols, Are Powerful Antioxidants Using an in Vitro Oxidation Model for Heart Disease. *Journal of Agricultural and Food Chemistry*, *43*(11), 2800–2802.

Viuda-Martos, M., Fernández-López, J., & Pérez-Álvarez, J. A. (2010). Pomegranate and its Many Functional Components as Related to Human Health: A Review. *Comprehensive Reviews in Food Science and Food Safety*, *9*(6), 635–654.

Wilcox, L. J., Borradaile, N. M., de Dreu, L. E., & Huff, M. W. (2001). Secretion of hepatocyte apoB is inhibited by the flavonoids, naringenin and hesperetin, via reduced activity and expression of ACAT2 and MTP. *Journal of Lipid Research*, *42*(5), 725–34.

Williams, R. J., Spencer, J. P., & Rice-Evans, C. (2004). Flavonoids: antioxidants or signalling molecules? *Free Radical Biology and Medicine*, *36*(7), 838–849.

Witko-Sarsat, V., Gausson, V., & Descamps-Latscha, B. (2003). Are advanced oxidation protein products potential uremic toxins? *Kidney International*, *63*(84), S11–S14.

Wojakowski, W., Gminski, J., Siemianowicz, K., Goss, M., & Machalski, M. (2001). The influence of angiotensin-converting enzyme inhibitors on the aorta elastin metabolism in diet-induced hypercholesterolaemia in rabbits. *Journal of the Renin-Angiotensin-Aldosterone System: JRAAS*, *2*(1), 37-42.

Yuan, T., Ma, H., Liu, W., Niesen, D. B., Shah, N., Crews, R., Rose, K. N., Vattem, D. A., Seeram, N. P. (2016). Pomegranate's Neuroprotective Effects against Alzheimer's Disease Are Mediated by Urolithins, Its Ellagitannin-Gut Microbial Derived Metabolites. *ACS Chemical Neuroscience*, *7*(1), 26–33.

Zand, R. S. R., Jenkins, D. J. A., & Diamandis, E. P. (2000). Steroid hormone activity of flavonoids and related compounds. *Breast Cancer Research and Treatment*, *62*(1), 35–49.

Zhao, X., Yuan, Z., Fang, Y., Yin, Y., & Feng, L. (2014). Flavonols and flavones changes in pomegranate (Punica granatum L.) fruit peel during fruit development. *Journal of Agricultural Science and Technology*, *16*, 1649–1659.

In: Polyphenolics
Editor: Patricia Clark

ISBN: 978-1-53610-709-8
© 2017 Nova Science Publishers, Inc.

Chapter 5

EFFECTS OF DIETARY POLYPHENOLS ON GENE EXPRESSION: EVIDENCE FROM EXPERIMENTAL STUDIES

Maria Rosana Ramirez, PhD
National Council of Scientific and Technical Research,
Faculty of Food Science, National University of Entre Rios,
Concordia, ER, Argentina

ABSTRACT

Worldwide mortality from infectious ailments is being substituted by chronic degenerative diseases, such as cancer, obesity, diabetes mellitus and neurological conditions. It is known that natural compounds may affect the expression and gene transcription which influence mechanisms involved in chronic diseases. Furthermore, they may also influence the predisposition to certain disorders as a result of individual genetic variability. For these reasons, nutrition research focuses not only on nutrient deficiency but also in the prevention of various progressive diseases. The inclusion of natural products such as fruit, beverages, teas and vegetables in the diet has long been related with various health benefits. These beneficial effects have been associated to the presence of polyphenolic compounds and their pharmacological properties. With regards to beneficial properties, some reports have revealed that polyphenolics could interact with cellular signaling pathways regulating the activity of transcription factors and affecting the expression of genes.

Genes differentially expressed include genes involved in a wide range of physiological and pathological functions, such as metabolism, transport, enzyme activity, signal transduction or transcription. In this chapter, we review studies assessing modulation of genes expression by dietary polyphenolics that could constitute a new pathway by which these phytocompounds may exert their health effects.

Keywords: gene expression, polyphenols

INTRODUCTION

Polyphenolics compounds represent a major group of secondary plant compounds and possess a great diversity with different structural and sterical properties. A broad spectrum of biochemical and pharmacological functions is attributed to these phytocompounds that might benefit human health [1, 2, 30]. Their protective activity in various experimental models of diseases *in vitro* and *in vivo* has been documented [3, 4]. Further, experimental evidence shows that this is partially attributable to changes in gene expression. Thus, these preventive agents are very promising in that they offer a non-toxic route to activate or repress particular gene expression pathways and, thereby to manage or prevent disease [4, 5, 9]. Here summarize key signaling events by which selected polyphenols can affect gene expression, in the context of some degenerative disease prevention.

1. MECHANISMS OF ACTION OF INDIVIDUALS POLYPHENOLS

1.1. Regulation of Genes Related to Inflammation

The inflammatory response is a natural process attributed to the defence response by the organism to a variety of external and detrimental stimuli [6]. The principal inflammatory stimuli are the pro-oxidants produced from activated neutrophils and macrophages have been informed to play a critical role in the pathogenesis of diverse inflammation-related diseases including diabetes, atherosclerosis, and cancer and neurodegenerative disorders. Polyphenols proficiently modulates the redox status and thus may play a critical role in regulating gene expression of inflammatory mediators in lipopolysacaride stimulated, RAW 264.7 macrophages [30, 52].

With regards to anti-inflammatory properties, some reports have revealed an anti-inflammatory effect of flavonol quercetin, for its ability to modulate the production and gene expression of the proinflammatory cytokine tumor necrosis factor alpha (TNFα) by human peripheral blood mononuclear cells (PBMC). Observations made from this experiment indicated that, quercetin significantly decrease TNF-α production and gene expression in a dose-dependent manner [38]. TNF-α plays a major role in modulation of the inflammatory response for example, stimulating secretion of interleukin-1 (IL-1), interleukin-6 (IL-6), platelet-derived growth factor and inducing expression of adhesion molecules.

Also, TNF-α promote the production of reactive oxygen and nitrogen species by leucocytes. This observation suggests that, the anti-inflammatory effect exerts by quercetin could be due to a decrease of pro-inflammatory cytokines TNFα via modulation of nuclear factor-κB (NF-κB) [40]. Nuclear factor κB is a transcription factor that plays a role in the gene regulation related with inflammation. Likewise, the NFκB signaling pathway is associated to the activation of mitogen-activated protein kinases (MAPKs), which stimulate transcription factors that promote inflammatory gene expression. These signaling pathways include c-Jun Nterminal kinase (JNK), p38MAP kinase (p38), and mitogenic signaling extracellular signal-regulated kinase 1/2 (ERK).

Resveratrol (trans-3,4',5-trihydroxystilbene) is a non-flavonoid polyphenolic found in grapes, some berries and red wine [3]. In the same way, studies have documented that resveratrol has various health benefits such as antioxidant, anti-inflammatory, antimutagenic, and anticarcinogenic effects [4]. That includes the inhibition of synthesis and release of pro-inflammatory mediators, modifications of eicosanoid synthesis, inhibition of some activated immune cells, or inhibiting the enzymes, such as cyclooxygenase (COX -1, COX-2), which are responsible for the synthesis of pro-inflammatory mediators through the inhibitory effect of resveratrol on transcription factors like nuclear factor kappaB (NF kappaB) or activator protein-1 (AP-1), and gene-regulatory functions [6, 7]. Although the precise mechanism of resveratrol action is not known yet, the resorcin moiety appears to be essential for the preventive efficacy [3, 4, 12, 32, 73].

Curcumin is a phytochemical constituent of turmeric (*Curcuma longa*) and is used as a spice to give a singular flavor and yellow color to Asian curries [5]. This phytocompound is derived from the rhizome, or root, of the

Curcuma plant. Curcumin has been demonstrated to have anti-inflammatory, antioxidative, antitumor and antidiabetic effects in a variety of *in vitro* or in *vivo* systems. With regards to anti-inflammatory properties, some reports have revealed an inhibitory effect of curcumin on mediators of inflammation as NFkappaB, lipoxygenase (LOX), cyclooxygenase-2, and inducible nitric oxide synthase (iNOS). These finding contributed to characterize the biological activity of *Curcuma longa* and to understand the ability of its extracts to enhance anti-inflammatory responses.

Luteolin, a plant flavonoid isolated from *Lonicera japonica* (Caprifoliaceae), has special anti-inflammatory properties both *in vitro* and *in vivo*. Chen et al. (2007) examined the potential anti-inflammatory properties of luteolin in LPS-stimulated mouse alveolar macrophage MH-S and peripheral macrophage RAW 264.7 cell lines [9]. Observations made from these experiments indicated that, luteolin can prevent the expression of inflammatory genes and accumulation of chemical mediators of inflammation such as nitric oxide (NO) and prostaglandin E2 (PGE2), as well as the expression of inducible NO synthase (iNOS), cyclooxygenase-2 (COX-2), tumor necrosis factoralpha (TNF-α), and interleukin-6 (IL-6) in mouse alveolar macrophage MH-S and peripheral macrophage RAW 264.7 cells.

Thus, luteolin is associated with the downregulated expression of iNOS and COX-2 as well as the attenuation of inflammatory cytokines, probably due to the suppression of NF-κB and AP-1 (activating protein-1) activation, resulting in its anti-inflammatory effects. Although in folk medicine, *Lonicera japonica* is widely used as anti-inflammatory. Based on these results the authors suggest that, the anti-inflammatory property of *L. japonica* may be partly due to the luteolin, providing support for the traditional use of this plant in folk medicine against alveolar inflammatory disorders [9].

Green tea is the most effective natural beverage for some disease prevention in humans. Green tea is rich in polyphenol compounds about 30% of dry weight of a leaf, with the main phytocompounds being epigallocatechin-gallate (EGCG), (-)-epigallocatechin (EGC), (-)-epicatechin (EC), and (-)-epicatechin-3-gallate (ECG). All of the phytocompounds have a wide variety of bioactivities relating to their chemical structure, but the different mechanisms underlying these actions have not been fully clarified [11].

Regardless of the mechanism of catechins they have long been recognized that it exhibits anti-inflammatory activity in addition to its antioxidant capacity. This could be due to their capacities to scavenge NO, the peroxynitrite anion, or to reduce the activity of NO synthase, including the neuronal nNOS and the inducible iNOS isoforms could also be inhibited by catechins [20, 21, 29]. This effect might be related to the inhibition of the activation of the transcription factor NF-κB as the κB sequence is present in the promoter region of the iNOS gene. Therefore it can be deduced that, these activities contribute to the anti-inflammatory effects of catechins [33, 46, 66, 67, 72]. It was also proposed another mechanism of action related to the presence of antioxidant response element (ARE) on iNOS gene promoter. Thus catechins could bind to the ARE and activate iNOS [15, 22].

Observations made from others experiments indicated that EGCG, was able to inhibit the production of IL-1 and attenuate the expression of cyclooxygenase-2 induced by IL-1 and amyloid β peptide (Aβ) or the activation of NF-κB and MAPK pathways induced by IL-1 and Aβ [40, 70].

On the other hands, Zhang et al. (2006) determined the protection of animal models with alcoholic liver damage and, evaluated gene expression of cytokines in alcoholic liver disease affected by tea polyphenol using cDNA chip technology [74]. For this, animal were divided into 3 groups, group A treated with alcohol for 12 weeks, group B fed alcohol plus polyphenol and control group (saline). After 12 weeks the rats were sacrificed. Different assays were performed showing that, the cytokine gene expression occurred in all groups. Moreover it was observed that, mRNA of IL3, IL4, IL-1R2, IL-6R and IL-7R2 was up-regulated after infusion of tea, but mRNA of IL-3Ra, IL-1R1 was down-regulated. Additionally inhibited the patophysiological changes of alcoholic liver disease.

Studies relating to chronic alcohol consumption have shown that, EtOH is a toxic substance which is readily absorbed in the gastrointestinal tract and mainly metabolized in the liver, causing metabolic imbalances, acetaldehyde production and free radicals. They conclude that the effect of the tea extract blocks the development of alcoholic liver disease by amelioration of inflammatory reaction and oxidative stress; and suggest that, the mechanism of action may be related to the regulation of gene expression of the cytokines secreted by hepatic cells and inflammatory cells [74, 73, 63].

1.2. Regulation of Genes Related to Chemoprevention

Chemoprevention by natural products has received growing attention in recent years as a promising approach in controlling the incidence and/or prevention of cancer. Some Heat-shock proteins (HSPs), or stress proteins, play roles in folding/unfolding of proteins, transport of proteins, cell-cycle control and signaling, and protection of cells against stress/apoptosis [13, 43,

71]. On the other hand, some reports have indicated that, members of the hsp70 and hsp27 families are highly expressed in breast and lung cancer and leukaemias and they play a role in the acquired resistance to chemotherapy treatment combined with hyperthermia [17].

More recently, hsps also known as chaperones, has been identified as a rational target for cancer diagnosis and therapy. Rusak et al. (2002) advanced on the studies and analyzed the effect of quercetin and five structurally related flavonoids (50 μM) on hsp 90α, hsp 70A, hsp 60 and hsp 27 gene expression in heat-stressed and heat-unstressed human promyeloid leukemia HL-60 cells. For this purpose, the effects of quercetin, kaempferol, methylquercetagetin, myricetin, taxifolin and isorhamnetin were investigated [57].

It was observed that, the hsp27 gene expression was inhibited by flavonoids more strongly than other hsp genes investigated in heat stressed as well as in unstressed cells. Based on these facts, it is possible that the position, number and substitution of hydroxyl groups of the B ring and saturation of the C2-C3 bond are important factors affecting flavonoid activity on hsp gene expression [26, 39].

Similarly, a number of studies carried out with quercetin (3,3′,4′,5,7-pentahydroxyflavon) and KNK437 (N-formyl-3,4-methylenedioxy-benzylidene-γ-butyrolactam), demonstrated their abilities to inhibit the accumulation of hsp mRNA and protein as well as the development of heat shock-induced thermotolerance in HeLa, colon cancer, squamous carcinoma and glioblastoma cultured cells [13, 17, 18, 39]. These studies also found that, both agents have different mechanisms of action: quercetin suppresses the cellular levels of heat shock transcription factor (HSF) while KNK437 inhibited binding activity HSF-HSE (heat shock element enhancer) [13, 43].

In this sense, the effect of quercetin and KNK437 (benzylidene lactam compound), on heat-induced heat shock protein (hsp) gene expression in epithelial cells of *Xenopus laevis* A6 kidney was analyzed. Researchers found that, both quercetin and KNK437 reduced the heat shock-induced accumulation of hsp30, hsp47 and hsp70 mRNA in *X. laevis* cultured cells. Although, compounds had no effect on the relative level of a non-heat shock protein mRNA, ef1α (elongation factor 1 alpha), in either control or heat shocked cells. The analysis of the results also revealed that quercetin partially inhibited HSP30 protein accumulation. This observation is supported on previous findings that show that, both quercetin and KNK437 inhibited heat shock factor activity resulting in a repression of hsp mRNA and protein accumulation in human cultured cells [35, 48, 71].

It has been shown that, epidermal growth factor receptor (EGFR/erbB1/ HER1) is one of the family of four erbB receptors. Activation of this factor is originated by binding of ligands, such as epidermal growth factor (EGF) and transforming growth factor-a. This results in the formation of homo- or hetero-dimers and activation of the receptor tyrosine kinase, which, consequently leading to signaling cascades and regulation of target gene expression. Overexpression and aberrant activation of EGFR and the EGF signal pathway is connected with neoplastic cell proliferation, stromal invasion, resistance to apoptosis and angiogenesis. Despite this, it has been shown that interruption of EGF signaling impairs tumor growth.

Recently it was reported that curcumin activated the peroxisome proliferator-activated receptor-gamma (PPARg) in Moser cells, a human colon cancer-derived cell line, leading to inhibition of cell growth by inhibiting tyrosine phosphorylation of EGFR and suppressing gene expression of EGFR and cyclin D1 (Chen and Xu, 2005). Other experiment was performed in order to confirm if a curcumin extract had potential chemopreventive effects. This study was done using human colon cáncer derived cells, including Moser, Caco-2 and HT-29 [8, 62]. Observation made from this experiment indicated that curcumin reduced the DNA-binding activity of the transcription factor Egr-1 to the curcumin response element.

Also, curcumin reduced the trans-activation activity of Egr-1 by suppressing egr-1 gene expression, which required interruption of the ERK signal pathway and reduction of the level of phosphorylation of Elk-1 and its activity [61, 69]. In conclusion, it was determined from this experiment that curcumin inhibited human colon cancer cell growth by suppressing gene expression of EGFR through reducing the trans-activation activity of Egr-1. [61, 62]. These results are relevant, considering the evidence suggesting that EGFR is a rational target for cancer therapy, including colorectal cancer.

Similarly, the flavonoid flavone contained in fruits and vegetables was identified as a potential apoptosis inducer in human colonic cancer cells. Investigators determined the anti-apoptotic effect of flavonoid flavone in HT-29 cells using proteomics and oligonucleotide array technologies. The data revealed numerous changes in transcript levels of genes related to signaling, transcription, cancer development but also to metabolism. The apoptosis inducing action of flavone was higher than that of the anti-tumor drug camptothecin [16].

The analysis also showed that several genes related to cell cycle such as Bcl-xL, Bak, NF-kB, COX-2, p21, cyclin B and cyclin E were altered in cells upon flavone exposure. Treatment of cells with flavones also alters the proteins involved in intermediary metabolism. Thus the anti-apoptotic effect exerted by flavones could be associated with changes in the flux of energetic substrates. In addition, proteome examination identified diverse heat shock proteins, annexins, and cytoskeletal caspase substrates as regulated by flavone. It is important to note that, these protein classes are known to play a role in apoptosis induction and execution.

Likewise, the high dietary intake of selenium or soybean isoflavones reduces prostate cancer risk, and affect androgen-regulated gene expression [27]. Recently, Legg et al. (2008) advanced on the studies and analyzed the combined effects of selenium and isoflavones on androgen-regulated gene expression in rat prostate. For this, healthy rats were exposed to diets containing an adequate (0.33-0.45 mg/kg diet) or high (3.33-3.45 mg/kg) concentration of selenium as semethyl seleno cysteine and a low (10 mg/kg) or high (600 mg/kg) level of isoflavones.

The authors showed that, high selenium intake reduced expression of the androgen receptor, Dhcr24 (24-dehydrocholesterol reductase), and Abcc4 (ATP-binding cassette sub-family C member 4). High isoflavone intake decreased expression of Facl3 (fatty acid CoA ligase 3), Gucy1α3 (guanylate cyclase alpha 3), and Akr1c9. In the case of Abcc4, the combination of high selenium/high isoflavones had a greater inhibitory effect than either dietary compound alone. They concluded that, combined intake of high selenium and high isoflavones may achieve a greater cancer-chemopreventive effect than either compound supplemented individually [27].

Chemopreventive effects of EGCG of green tea may be based on several mechanisms. These include the inhibition of AP-1, the down-regulation of the NFκB-inducing kinase, the inhibition of HER2/Neu signaling (human epidermal growth factor receptor 2), or Vascular Endothelial Growth Factor (VEGF) production [14, 20]. Liver and small intestine gene expression profiles obtained from wild-type and Nrf2 knockout mice treated with EGCG or not unraveled the regulation of more than 600 genes by EGCG via Nrf2. Validation of the identified genes will elucidate a potential role of Nrf2 in the functions of EGCG.

More recent data point to an inhibition of the β-catenin/TCF transcriptional activity. A reduction of the expression level of β-catenin by physiological EGCG concentrations is achieved by an activation of the lysosomal trafficking of β-catenin and subsequent degradation [22, 29, 66, 67]. However, green tea extract have a dual effect in experimental carcinogenesis models: one protective activity at high concentrations and citotoxic activity at low concentration. DNA microarray analyses were performed to ascertain the gene expression profiles elicited by isothiocynates and green tea polyphenols at low concentrations as well as at higher concentrations [47, 70, 72].

It was determined from this experiment that, a low concentration these compounds activate mitogen-activated protein kinase (MAPK) pathway leading to activation of Nrf2 and antioxidant response element (ARE) with subsequent homeostatic response. At higher concentrations, these agents activate the caspase pathway, leading to apoptosis a potential cytotoxic effect if it occurred in normal cells. Based on these facts, it is possible that the studies of these signaling pathways may yield important insights during pharmaceutical drug discovery and development [22].

Apotosis plays important roles in diverse biological processes including carcinogenesis. In the human neuroblastoma SH-SY5Y cell line, epigallocatechin-gallate (EGCG) modified the expression of pro-apoptotic genes Bax and Bad while inducing the anti-apoptotic genes Bcl-2, Bcl-w, and Bcl-X in 6-hydroxydopamine-induced apoptosis. Furthermore, catechins may affect the process of apoptosis by altering the expression of anti-apoptotic and proapoptotic genes [70, 72].

After treating the transformed human bronchial epithelial 21BES cells with EGCG in presence or absence of catalase, was observed that: first, gene expression changes that are mediated by H_2O_2 including genes of the transforming growth factor-h signaling pathway, SMAD3, and TSC22. Second, induces gene expression changes that were not affected by catalase included those of the bone morphogenetic protein signaling pathway, peptidylprolyl isomerase (cyclophilin)–like 2, alkylated DNA repair enzyme alkB, polyhomeotic-like 2, and homeobox D1. However, the end result is a clear anticancer effect [29, 47].

In agreement with this, investigators demonstrated a protective effect of regular green tea consumption on the risk of developing breast cancer that is limited to women possessing the high-activity genotype of the angiotensin-converting enzyme gene. They found that, green tea polyphenols may exert their chemopreventive effect through an angiotensin II-driven pathway. It is known that, angiotensin II is a potent angiogenic factor and may upregulate NADPH oxidase in endothelial cells in a dose-dependent manner. The ROS derived from endothelial NADPH oxidase participates in vascular endothelial growth factor (VEGF) signaling and plays a role in VEGF-induced angiogenesis.

It has been suggested that, the reduction of angiotensin II levels by ACE inhibitors suppresses VEGF-induced angiogenesis and inhibits tumor growth in vivo. Green tea have been shown to decrease the production of ROS generated via the NADPH oxidase-dependent pathway, and reduce the level of VEGF secreted by human breast cancer cells in a dose dependent manner. Based on these finding, it is possible that green tea reduces breast cancer risk by inhibiting the angiotensin II/NADPH oxidase-induced ROS/VEGF pathway in breast cancer. These authors demonstrated for the first time epidemiological evidence of an interaction effect between green tea intake and angiotensin converting enzyme gene polymorphisms on risk of female breast cancer [72].

1.3. Regulation of Genes Related to Aging

Several human and animal studies show that, foods containing phytocompounds have the ability to reduce the risk of several degenerative diseases that frequently accompany aging. Consequently natural compounds mostly as a result of their antioxidant property can be beneficial in the prevention of age-related diseases [24, 30, 51]. In order to confirm this hypothesis, eleven classes of polyphenols commonly present in the diet were tested on the expression of genes important in cardiovascular health (endothelial NO synthase- eNOS), endothelin-1 (ET-1) and vascular endothelial growth factor (VEGF) by cultured vascular endothelial cells (human umbilical vein endothelial cells-HUVEC) under stress-free and oxidative stress conditions (in the absence or presence of H_2O_2) [42, 64].

It was observed that, resveratrol and quercetin increased eNOS mRNA expression and decreased ET-1 mRNA expression in HUVEC in a dose-dependent manner in the absence and presence of H_2O_2. Additionally resveratrol and quercetin decreased endothelin secretion, in the absence and presence of H_2O_2 (50 mM). In the same way, epigallocatechin gallate had similar effects on both the eNOS and ET-1 mRNA expression. However none of the other tested polyphenols produces similar effects. The authors concluded that, the effects on gene expression can result in vasodilation and decreased blood pressure [42, 64].

In the same way, the dietary isoflavone genistein has positive effects with regard to chronic diseases, such as cancer and cardiovascular disease, particularly on aspects related to blood pressure and angiogenesis. It has been reported that angiogenesis is partly regulated by redox sensing transcription factors and the expression of genes associated with angiogenic response, so it is likely to be regulated by antioxidants such as genistein. For this reason, the analysis of differential gene expression in the endothelium is essential to the understanding of the events leading to the formation of high blood pressure, atherosclerotic lesions, smooth muscle proliferation, and angiogenesis. Some studies, analyzed the gene expression profile of selected families of genes known to play an important role in molecular models of atherosclerosis using endothelial cell (HUVEC) [2, 42, 64, 65].

It was observed that, the treatment of HUVEC with genistein resulted in a significant down-regulation of genes encoding for endothelin-2 and endothelin-converting enzyme in basal conditions. This protein is member of a family of four isoforms, ET-1, ET-2, ET-3 and ET-4 that are the most powerful vasoconstrictors known, with ET-1 being the predominant form. Consequently, the down-regulation of endothelin-converting enzyme-1 and endothelin-2 by genistein, as observed in the current study, may lead to vasodilatation, and decrease in blood pressure as well as in the inhibition in inflammation, cell adhesion, platelet aggregation and chemotaxis. The authors deduced that physiologically achievable levels of genisteina affect the expression of mRNA encoding for proteins involved in the control of blood [2, 64, 65].

In agreement, other studies detected that treatment with flavone affects the genes involved in the signaling processes. For example, the MAP-kinase interacting Ser/Thr kinase-1, was augmented at the mRNA-level and a significant upregulation of erbin, an erb-B2 interacting protein that inhibits the MAP-kinase pathway was also detected suggesting an inhibition of the MAP-kinase pathway by flavone. These finding could be related to the effect on angiogenesis. However, most mRNA-species changed by flavone could be related to metabolism, confirming the strong metabolic adaptations of HT-29 cells in response to flavone treatment.

It is believed that the origin of atherosclerosis is due to the interaction between lipid metabolism, cytokines and cellular activity within the arterial wall [21]. Therefore, the genes encoding the LDL (LDL-R), LXR-α, CD36, PPAR (α-γ) and C-myc receptor genes play a crucial role in the interaction of these processes responsible for atherogenesis. In this regards, polyphenols extracts from green tea (*Camellia sinensis*, Theaceae) were investigated on transcriptional regulation of the genes involved in the genomics of atherosclerosis on human mononuclear cells [15, 22].

This experiment showed that extract induces an increase in the transcriptional expression of genes coding for PPARs (α γ) and LDL receptor (LDL-R), and this was accompanied by dose-dependent decrease in the expression of genes coding for PPAR (α γ), LXRα and CD-36. The dual effects found are relevant, considering the folk use of this green tea to prevent the atherosclerotic process. The authors concluded that, the inhibitory effect of green tea polyphenols on the development of atherosclerotic lesions can be explained by their ability to inhibit oxidation of LDL and, down-regulate genes coding for PPAR-γ, CD-36, LXR-α, C-myc coupled with up-regulation of genes coding for LDL-R and PPAR-α. Besides PPAR-activation has been shown to inhibit neutral lipid accumulation in macrophages thereby contributing to a regression of atherosclerosis [21, 30, 42, 64].

1.4. Regulation of Genes Related to Metabolic Disesase

Metabolic diseases are characterized by including: insulin resistance, hyperinsulenemia, obesity, dyslipidaemia, impaired glucose tolerance, hypertension and atherosclerosis [10]. All the mentioned features are associated with the development of both diabetes and cardiovascular disease (CVD) [28]. For instance, mucopolysaccharidoses (MPSs) are autosomal, or X-linked (type II) recessive lysosomal storage disorders caused by the deficiency in activity of a lysosomal enzyme involved in catabolism of glycosaminoglycans. Accumulation of glycosaminoglycans leads to severe clinical symptoms and significantly shortened life span because of damage of affected tissues and organs, including the heart, respiratory system, bones and joints, in most MPS types and subtypes, also central nervous system (CNS) [37].

Therapies used in MPS patients, such as cells' transplantations and enzyme replacement therapy, are ineffective for neurological symptoms, due to the poor distribution of enzyme in the central nervous system. Thus one strategy can be the application of the implementation of the non-enzymatic substrate reduction therapy using glycosaminoglycans metabolism modulators, such as various phytocompounds [38].

Phytochemicals were found previously to modulate efficiency of synthesis of glycosaminoglycans, compounds which are accumulated in cells of patients suffering from mucopolysaccharidoses. Recently investigators have utilized microarray technology to investigate the effects of phytocompounds: genistein, kaempferol, daidzein, used alone or in combinations, on expression of genes coding for proteins involved in glycosaminoglycans metabolism and lysosome biogenesis.

The results revealed that genistein, kaempferol and combination of these two compounds induced dose and time dependent remarkable alterations in transcript profiles of glycosaminoglycans metabolism genes in cultures of wild-type human dermal fibroblasts (HDFa). Moreover effects of the combination of genistein and kaempferol were stronger than those revealed by any of these phytocompounds used alone. Genistein, kaempferol and mixture of these phytocompounds also stimulated expression of transcription factor EB (TFEB). Additionally, a decrease in mTOR (mammalian Target of Rapamycin) transcript level was detected at these conditions. Therefore, understanding the mechanism of correction of cellular glycosaminoglycans levels by these components may contribute to potential application of them as possible drugs for mucopolysaccharidoses [2, 10, 28, 37].

2. ACTIVITIES OF POLYPHENOL EXTRACTS FROM FOOD

2.1. Cocoa Polyphenol Extract

Recently, incidence of obesity in human has increased becoming a prevalent public health problem. Obesity can predispose to a number of harmful disorders which can be accompanied by inflammation. It has been suggested that, Cocoa polyphenol (CP), due to their biological properties, may be supplementary treatments for adipose tissue-fat gain [1, 15].

Ali et al. (2015) investigated the hypothesis that, Cocoa Polyphenols treatment modulates expressing of lipid metabolism genes in mesenteric white adipose tissue (MES-WAT). Rodents were separated into various groups that were fed different diets (low-fat or high-fat) for 12 weeks. After that the animals were supplemented with Cocoa Polyphenols (600 mg/kg bw/day) extract for 4 weeks. It was observed that CP treatment significantly decreased gene expression levels for lipogenic enzymes, while increased the mRNA levels responsible for lipolysis enzymes. Under these conditions, these findings reveal a different insight into the molecular mechanisms underlying the physiological effect of Cocoa Polyphenols on obesity biomarkers in obese rodents [1, 15].

2.2. Blackcurrant Extract

It has been shown that, transintestinal cholesterol efflux (TICE) provides a non-biliary route for cholesterol excretion. Therefore it can be deduced that TICE activation can lower plasma cholesterol levels. To corroborate this hypothesis, Kim et al. (2013) investigated the effect of blackcurrant extract

(BCE) containing 25% polyphenols on the regulation of genes involved in intestinal cholesterol metabolism with a specific focus on transintestinal cholesterol efflux (TICE) [23].

Caco-2 cells, a human intestinal cell line, were incubated with BCE (50 or 100 μg/mL) for 24 h for gene expression analysis. It was observed that the extract has a dual effect, on the one hand BCE significantly decreased the expression of genes for cholesterol synthesis (3-hydroxy-3-methylglutaryl coenzyme A reductase and sterol regulatory element binding protein 2), for apical cholesterol uptake (Niemann-Pick C1 Like 1), and for basolateral cholesterol efflux (ATP binding cassette transporter (ABC) A1).

On the other, BCE significantly increased mRNA and protein levels of ABCG5/G8, which facilitate apical cholesterol efflux to the intestinal lumen. The authors concluded that, blackcurrant extract may activate TICE pathway by increasing LDL uptake into enterocytes and by facilitating the efflux of LDL-derived cholesterol to the luminal side via increased ABCG5/G8 expression in the intestine [23, 53].

2.3. Apples Extract

Apples like other fruits are a significant part of the world diet, and they are a good source of poliphenolics. These phytocompounds are the bioactive metabolites present in the fruit. The polyphenols and flavonoids that were identified in the fruit are epicatechin, phloridzin, chlorogenic acid, caffeic acid, quercetin-3-rutinoside, quercetin-3-galactoside, quercetin-3-glucoside, quercetin-3-rhamnoside [68].

A recent study examined the phytocompounds from apple extract (AE), to determine if they contribute to reducing risks during colon carcinogenesis by inhibiting tumor cell growth or by favorably modulating expression of drug metabolism genes. The authors showed that, AE strongly inhibited growth of HT29 cells and markedly influenced expression of genes encoding xenobiotic enzymes in subtoxic concentrations [68].

Treatment with AE resulted in an upregulation of several genes such as: GSTP1, GSTT2, MGST2 (Phase II, glutathione S-transferases), cytochrome P450, family 4, subfamily F, polypeptide 3 (CYP4F3, P450 gene family-Phase I), carbohydrate sulfotransferase 5, 6, 7 (CHST5, CHST6, CHST7, sulfotransferases-Phase II); and downregulation of Phase II epoxide hydrolase gene (EPHX1), in relation to the controls. This observation suggests that these genes could be new targets for chemoprevention.

Others studies, was also found an inhibitory effect of the polyphenol enriched apple juice extract AE04 on inflammatory gene expression in LPS/IFN-c stimulated Mono-Mac6 cells [19]. The same researchers recently analyzed the preventive effectiveness of polyphenolic juice extracts and single major constituent's procyanidin B1, procyanidin B2, phloretin, and phloridzin on the expression of pro-inflammatory marker genes in human immune relevant and epithelial colon carcinoma cell lines using qRTPCR.

In this study, the effect on NF-κB, IP-10-, IL-8-promoter, STAT1-dependent signal transduction, and the relative protein levels of multiple released cytokines and chemokines were also examined. Moreover it was observed that, the apple juice extract AE04 significantly inhibited the expression of NF-jB regulated proinflammatory genes (TNF-α, IL-1b, CXCL9, CXCL10), inflammatory relevant enzymes (COX-2, CYP3A4), and transcription factors (STAT1, IRF1) in LPS/IFN-c stimulated MonoMac6 cells without significant effects on the expression of house-keeping genes.

They also observed that, procyanidin B2 is mainly responsible for the anti-inflammatory activity of AE04. Phloretin and procyanidin B1 inhibited proinflammatory gene expression and suppressed NF-jB, IP-10, IL-8-promoter, and STAT1-dependent signal transduction in a dose-dependent manner. These finding contributed to characterize the biological activity of Apple and to understand the ability of its extracts to enhance inflammatory responses [19, 68].

2.4. Berries Extract

As with other fruits, berries contain a range of micronutrients and secondary metabolites which are important for health, such as phenolic acids as well as anthocyanins, proanthocyanidins and other flavonoids [51-53]. Recently, Shafiee-Kermani et al. (2009) examine the effect of lower concentration range of blueberry phenolic extracts (6.25-100 µg/ml) upon cell proliferation, oxidative metabolism, and gene expression related to cell-cycle progression, and the epigenetic machinery, using HepG2 cells as an *in vitro* model of hepatocarcinoma [60].

Different assays were performed showing that the treatment of human hepatocarcinoma, HepG2 induces changes in gene expression. These genes are related cell-cycle regulation (cyclin D1, cyclin-dependent kinase inhibitor 1A, and proliferating cell nuclear antigen, *PCNA*), antioxidant metabolism (glutamate-cysteine ligase catalytic subunit and glutathione reductase), and epigenetic machinery related to cell-cycle progression (DNA-methyltransferase 1, DNA-methyltransferase 3a, and Sirtuin 1). Besides, these results also support the popular use of blueberry in gastric disturbances.

Epidemiological, animal, and other studies have shown that, supplementation with berries are effective in reducing oxidative stress associated with aging and, are beneficial in reversing age-related neuronal and behavioral changes [51, 52]. These findings are consistent with the hypothesis that phytocompounds can have effects in cell signaling and decrease oxidative damage, and also suggest that they might act directly on gene expression in animals [53-55].

In that sense, researchers evaluated the effect of chronic supplementation (orally) with strawberry guava fruit (*Psidium catleyanum*), on the expression in mice hippocampus using microarrays. It was observed that 38 genes encoding for signal transduction and 29 genes related to protein synthesis (6 genes for protein degradation) were up or down-regulated by *Psidium cattleyanum* (Myrtaceae) extract [55]. These included ribosomal proteins, Serpinb1a, PEX1, PGAP1, DIP2c and Usp47, a protein associated with regulation of many essential cellular processes usually by the degradation of regulators of these processes such as cell cycle, proliferation, differentiation, and signal transduction among others.

Taken together, analysis of gene expression might be useful to elucidate the mechanisms of phytochemical action and may also help to identify potential targets for further examination. It was deduced that, these effects may contribute to the prevention of age-related and pathological degenerative processes in the brain; however, the effects of these phytocompounds in these pathological conditions remain to be verified.

2.5. Red Wine

In recent years there has been a significant increase in the consumption of wine by the human population. This is most likely a result of recently acknowledged health benefits that this beverage possesses. Red wine have been shown to have healing properties, to block carcinogenesis and to inhibit the growth of tumors in animals, or in cell culture by altering the activity of certain enzymes or the expression of specific genes [3, 60, 73].

Soleas et al. (2001) studied the anticancer properties of red wine and its polyphenolic constituents as well as their ability to modulate the expression of p53 gene expression in cancer cells. For this experiment, they used 3 three human breast cancer cell lines (MCF-7, T47D; MDA-MB-486) and one human colon cancer cell line (Colon 320 HSR 1). In the first step, cell lines were for 24-h with each of four polyphenols: quercetin (flavone), catechin (flavan-3-ol), *trans*-resveratrol (trihydroxystilbene), caffeic acid (hydroxycinnamate), at concentrations ranging from 1027 M to 1024 M [48, 50]. In the second step, p53 concentrations were measured in cell lysates by a time-resolved fluorescence immunoassay.

It is known that, p53 gene has been connected in normal cell proliferation, cell cycle control, induction of apoptosis, DNA repair, and carcinogénesis. The product of this gene controls transcriptional activation of other genes that, cumulatively lead to cell cycle arrest and apoptosis. Therefore, mutations in the p53 gene are crucial for the transition of cells from normal to malignant phenotype. These authors demonstrated that, treatment does not affect uniformly the expression of p53 protein in the four cell lines studied. The changes with all of these phytocompounds were modest and none of these effects was dose-dependent. Therefore, the anti-carcinogenic activity of wine should not be attributed to modulation of p53 gene expression [59, 63, 73].

2.6. Seeds Polyphenolic Extract

In order to evaluate uses of specific polyphenolics in the reduction of stress oxidative, the effects of seed extracts of genotypes from *Arabidopsis thaliana* were examined in animals. Extracts from this plant has been known to possess therapeutic properties and consequently de phytocompounds in these plant genotypes were used in one study. For this, investigators administered seeds of different flavonoid contrasted genotypes of *Arabidopsis thaliana* to rats, mixed with their diet, with the objective of determining the potential effects on colon and liver DNA oxidative stress and gene expression profiles [34].

They divided the animals into groups to distinguish among flavonols alone, flavonols with proanthocyanidins and flavonols with anthocyanins (15% w/w seeds for 4 weeks). For this the following seeds were used: Ws-2 wild-type containing flavonols and PAs, tt3-4 mutant containing flavonols only, ban-5 accumulating flavonols and anthocyanins, tt4-8 mutant, deprived of flavonoids.

DNA oxidative damage was significantly reduced only in the liver of rats fed tt3-4 mutant seeds. Under these conditions, microarray analysis of the liver revealed down-regulation of genes associated with oxidative stress, Krebs cycle, electron transport and proteasome degradation in all treated groups compared to the tt4-8-fed control rats. These effects could be due to a number of polyphenols including flavonoid content.

By the other hand, was observed a down-regulation of inflammatory response genes in the colon mucosa in ban-5-fed rats. This experiment suggests that, this effect may be related to the content of anthocyanin [34]. In conclusion this study proposes that a method of reducing the DNA oxidative damage in rats could be through the addition of dietary *Arabidopsis thaliana* seeds supplementation.

2.7. Oliva Extract

Olive oil contain a large array of phenolic constituents that have been associated with some beneficial health effects in rats and human, such as delayed mortality, and decreased oxidative damage in blood, or the expression of specific genes. An experiment was performed in order to determine if an extract produced from olive oil had antioxidant effect during the course of liver regeneration following one-third hepatectomy in mice [36].

They also evaluated the changes in the redox state-regulating enzymes, thiol proteins and Nrf2 gene signatures that coordinate adaptive stress response. For this purpose, mice were intraperitoneally given either oil extract (50mg/kg bw) or saline for seven consecutive days, while respective controls received vehicle alone. The authors demonstrated the efficacy of pre-exposure to olive oil polyphenols extract in stimulating liver regeneration in an experimental model, and this effect was associated with an increase of Nrf2 mRNA expression level. They also showed that, olive oil polyphenols can act as oxidants in some circumstances and activate defensive mechanisms under the control of redox-sensitive genes which could have stimulating beneficial health reaction during stressful conditions.

Other studies were performed to determine whether polyphenols of olive oil could exert an *in vivo* effect on genes related to cholesterol efflux in humans [49]. Thus, in a randomized, controlled, cross-over trial, 13 pre/hypertensive populations were assigned 30 ml of two similar olive oils with high (961 mg/kg) and moderate (289 mg/kg) polyphenol content. The study revealed, an up-regulation of the expression of cholesterol efflux related genes ABCA1, SR-B1, PPARBP, PPARα, PPARγ, PPARδ, CD-36 and COX-1 in human white blood cells occurs after a polyphenol-rich olive oil ingestion versus a moderate one. The results also showed that changes in gene expression were related to a decrease in oxidized low-density lipoproteins and, with an increase in oxygen radical absorbance capacity and olive oil polyphenols. These results support a significant role of olive oil polyphenols in the up-regulation of genes involved in the cholesterol efflux from cells to HDL in vivo in humans [36].

In the same way, it is known that long-term treatment with extra-virgin olive oil with phenolic compounds was related with beneficial health effects in ageing rats, such as delayed mortality, reduced incidence of ulcerative dermatitis, and decreased oxidative damage in blood. In a study authors select an extra-virgin Italian olive oil rich in phenolic compounds administered to rats and middle-aged. Further, the oil was mixed with the animal food to reach a final fat proportion of 23% (w/w, dry diet), and mimic the total lipid contribution common in the diet of humans living in Western Europe and North America [49]. They used as a control, an identical olive oil with a lower content of antioxidants. Moreover, it was administered maize oil as an additional control to test the effects of different antioxidants and fatty acids.

After treatment, were evaluated a series of biochemical parameters related to oxidative stress in the brain and functional tests to evaluate motor, cognitive, emotional behaviour and pain sensitivity. It was observed that, treatment does not exert significant protective actions on motor and cognitive functions of the aging brain, but they might have an effect on anxiety-associated behaviour, possibly through a modulation of the expression level of glutathione reductase, a gene involved in the defense mechanisms against oxidative stress [31, 49].

CONCLUSION

Polyphenols compounds appear to be potential agents against diseases that occur in an age dependent manner and proceed chronically, due to their ability to influence and modulate various cellular processes such as signaling, proliferation, apoptosis, differentiation and stress response. Experimental evidence also suggests that this is partially attributable to changes in gene expression. In fact, consumption of a diet rich in polyphenolics is associated with beneficial effects to animal and human health. However it is unwise to extrapolate these results to the human situation without appropriate clinical trials; because there is still much to learn in terms of correlative versus causal effects of human exposure to polyphenol compounds. For these reasons, pharmacological studies referring to pharmacodynamic and pharmacokinetic should be performed in animals and humans.

REFERENCES

[1] Ali, F., Isamil, A., Esa, M.N., Pei, P.C. 2015. Transcriptomics expression analysis to unveil the molecular mechanisms underlying the cocoa polyphenol treatment in diet-induced obesity rats. *Genomics.* 105(1), 23-30.

[2] Ambra, R., Rimbach, G., S. De Pascual T., Fuchs, D., Wenzel, U., Daniel, H., Virgili F. 2006. Genistein affects the expression of genes involved in blood pressure regulation and angiogenesis in primary human endothelial cells. *Nutr. Metab. Cardiovasc.* 16(1), 35-43.

[3] Bastianetto, S., Zheng, W. H., Quirion, R. 2000. Neuroprotective abilities of resveratrol and other red wine constituents against nitric oxide-related toxicity in cultured hippocampal neurons. *Br. J. Pharmacol.* 131, 711-720.

[4] Baur, J. A. and Sinclair, D. A. 2006. Therapeutic potential of resveratrol: The in vivo evidence. *Nat. Rev. Drug. Discov.* 5, 493-506.

[5] Bengmark, S. 2006. Curcumin, an atoxic antioxidant and natural NFκB, cyclooxygenase-2, lipooxygenase, and inducible nitric oxide synthase inhibitor: a shield against acute and chronic diseases. *Parenter. Enteral. Nutr.* 30, 45-51.

[6] de la Lastra, C.A., Villegas, I. 2005. Resveratrol as an anti-inflammatory and anti-aging agent: mechanisms and clinical implications. *Mol. Nutr. Food Res.* 49, 405-430.

[7] Chao, H.H., Juan, S.H., Liu, J.C., Yang, H.Y., Yang, E., Cheng, T.H., Shyu, K.G. 2005. Resveratrol inhibits angiotensin II-induced endothelin-1 gene expression and subsequent proliferation in rat aortic smooth muscle cells. *Eur. J. Pharmacol.* 515, 1-9.

[8] Chen, A., Xu, J., Johnson, A.C. 2006. Curcumin inhibits human colon cancer cell growth by suppressing gene expression of epidermal growth factor receptor through reducing the activity of the transcription factor Egr-1. *Oncogene.* 25, 278-287.

[9] Chen, C., Peng, W., Tsai, K., Hsu, S. 2007. Luteolin suppresses inflammation associated gene expression by blocking NF-κb and AP-1 activation pathway in mouse alveolar macrophages. *Life Sci.* 81, 1602-1614.

[10] Eckel, R. H., Grundy, S.M., Zimmet, P.Z. 2005. The metabolic syndrome. *Lancet.* 365, 1415-1428.

[11] Graham, H.N. 1992. Green tea composition, consumption, and polyphenol chemistry. *Prev. Med.* 21, 334-350.

[12] Han, Y. S., Zheng, W. H., Bastianetto, S., Chabot, J.G., Quirion, R. 2005. Neuroprotective effects of resveratrol against beta-amyloid induced neurotoxicity in rat hippocampal neurons: involvement of protein kinase C. *Br. J. Pharmacol.* 141, 997-1005.

[13] Hansen, R. K., Oesterreich, S., Lemieux, P., Sarge, K.D., Fuqua, S.A.W. 1997. Quercetin inhibits heat shock protein induction but not heat shock factor DNA-binding in human breast carcinoma cells. *Biochem. Biophys. Res. Commun.* 239, 851-857.

[14] Hayes, J. D., McMahon, M. 2001. Molecular basis for the contribution of the antioxidant responsive element to cancer chemoprevention. *Cancer Lett.* 174, 103-113.

[15] Heo, H.J., Lee, C.Y. 2005. Epicatechin and catechin in cocoa inhibit amyloid beta protein induced apoptosis. *J. Agric. Food Chem.* 53, 1445-1448.

[16] Herzog, A., Kindermann, B., Döring, F., Daniel, H., Wenzel, U. 2004. Pleiotropic molecular effects of the pro-apoptotic dietary constituent flavone in human colon cancer cells identified by protein and mRNA expression profiling. *Proteomics.* 4, 2455-2464.

[17] Hosokawa, N., Hirayoshi, K., Kudo, H., Takechi, H., Aoike, A., Kawai, K., Nagata, K. 1992. Inhibition of heat shock factor *in vivo* and *in vitro* by flavonoids. *Mol. Cell. Biol.* 12, 3490-3498.

[18] Jakubowicz-Gil, J., Rzymowska, J., Gawron, A. 2002. Quercetin, apoptosis, heat shock. *Biochem. Pharmacol.* 62, 1591-1595.

[19] Jung, M., Triebel, S., Anke, T., Richling, E., Erkel, G. 2009. Influence of apple polyphenols on inflammatory gene expression. *Mol. Nutr. Food Res.* 53, 1263-1280.

[20] Kato, K., Long, N.K., Makita, H., Toida, M., Yamashita, T., Hatakeyama, D., Hara, A., Mori, H. and Shibata, T. 2008. Effects of green tea polyphenol on methylation status of RECK gene and cancer cell invasion in oral squamous cell carcinoma cells. *Br. J. Cancer.* 99, 647-654.

[21] Kaul, D., Sikand, K., Shukla A. R. 2004. Effect of green tea polyphenols on the genes with atherosclerotic potential. *Phytother. Res.* 18, 177-179.

[22] Kim, S. J., Jeong, H. J., Lee, K. M., Myung, N. Y., An, N. H., Mo Yang, W., Kyu Park, S., Lee, H. J., Hong, S. H., Kim, H. M., Um, J. Y. 2007. Epigallocatechin-3-gallate suppresses NF-κB activation and phosphorylation of p38 MAPK and JNK in human astrocytoma U373MG cells. *J. Nutr. Biochem.* 32 (10), 1720-1725.

[23] Kim, B., Perkins, A., Lee, J. 2013. Regulation of genes involved intestinal Cholesterol Metabolism by Polyphenol-Rich Black Currant Extract in Caco-2 Cells. *FASEB J.* 27, Supplement 1078.8.

[24] Kluth, D., Banning, A., Paur, I., Blomhoff, R., Brigelius-Flohé, R. 2007. Modulation of pregnane X receptor-and electrophile responsive element-mediated gene expression by dietary polyphenolic compounds. *Free Radic. Biol. Med.* 42, 315-325.

[25] Koishi, M., Hosokawa, N., Sato, M., Nakai, A., Hirayoshi, K., Hiraoka, M., Abe, M., Nagata, K. 1992. Quercetin, an inhibitor of heat shock protein synthesis, inhibits the acquisition of thermotolerance in a human colon carcinoma cell line. *Jpn. J. Cancer Res.* 83, 1216-1222.

[26] Lee, Y. J., Erdo, G., Hou, Z., Kim, S.H., Kim, J.H., Cho, J.M., Corry, P.M. 1994. Mechanism of quercetin-induced suppression and delay of heat shock gene expression and thermotolerance development in HT-29 cells. *Mol. Cell. Biochem.* 137, 141-154.

[27] Legg, R. L., Tolman, J. R., Lovinger, C. T., Lephart, E. D., Setchell, K., Christensen, M.J. 2008. Diets high in selenium and isoflavones decrease androgen-regulated gene expression in healthy rat dorsolateral prostate. *Reprod. Biol. Endocrinol.* 6, 57-64.

[28] Levitan, E. B., Song, Y., Ford, E. S., Liu, S. 2004. Is nondiabetic hyperglycemia a risk factor for cardiovascular disease? A meta-analysis of prospective studies. *Arch. Intern. Med.* 164, 2147-2155.

[29] Levites, Y., Youdim, M.B., Maor, G., Mandel, S. 2002. Attenuation of 6-hydroxydopamine (6-OHDA)-induced nuclear factor-κB (NF-κB) activation and cell death by tea extracts in neuronal cultures. *Biochem. Pharmacol.* 63, 21-29.

[30] Li, B., Xiong, M., Zhang, H. 2014. Elucidating Polypharmacological Mechanisms of Polyphenols by Gene Module Profile Analysis. *Int. J. Mol. Sci.* 15, 11245-11254.

[31] Li, S., Lin, J. D. 2009. Molecular control of circadian metabolic rhythms. *J. Appl. Physiol.* 107, 1959-1964.

[32] Liu, J.C., Chen, J. J., Chan, P. 2003. Inhibition of cyclic strain-induced endothelin-1 gene expression by resveratrol. *Hypertension* 42, 1198-1205.

[33] Lorenz, M., Wessler, S., Follmann, E., Michaelis, W., Dusterhoft, T., Baumann, G., Stangl, K., Stangl, V. 2004. A constituent of green tea, epigallocatechin-3-gallate, activates endothelial nitric oxide synthase by a phosphatidylinositol-3-OH-kinase-, cAMP-dependent protein kinase-, and Akt-dependent pathway and leads to endothelial-dependent vasorelaxation. *J. Biol. Chem.* 279, 6190-6195.

[34] Luceri, C., Giovannelli, L., Pitozzi, V., Toti, S., Castagnini, C., Routaboul, J., Lepiniec, L., Larrosa, M., Dolara, P. 2008. Liver and colon DNA oxidative damage and gene expression profiles of rats fed Arabidopsis thaliana mutant seeds containing contrasted flavonoids. *Food Chem. Toxicol.* 46, 1213-1220.

[35] Manwell, L. A., Heikkila, J. J. 2007. Examination of KNK437 and quercetin mediated inhibition of heat shock-induced heat shock protein gene expression in Xenopus laevis cultured cells. *Comp. Biochem. Physiol. A.* 148, 521-530.

[36] Marinic, J., Brozni, D., Milin, H. 2015. Preexposure to olive oil polyphenols extract increases oxidative load and improves liver mass restoration after hepatectomy in mice via stress-sensitive genes. *Oxid. Med. Cell. Longev.* 2016, 1-13.

[37] Miranda, P. J., Defronzo, R. A., Califf, R. M., Guyton, J. R. 2005. Metabolic syndrome: definition, pathophysiology, and mechanisms. *Am. Heart J.* 149, 33-45.

[38] Myhrstad, M. C. W., Carlsen, H., Nordstrom, O., Blomhoff, R., Moskaug, J. O. 2002. Flavonoids increase the intracellular glutathione level by transactivation of the γ-glutamylcysteine synthetase catalytical subunit promoter. *Free Radic. Biol. and Med.* 32, 386-393.

[39] Nagai, N., Nakai, A., Nagata, K. 1995. Quercetin suppresses heat shock response by down regulation of HSF1. *Biochem. Biophys. Res. Commun.* 208, 1099-1105.

[40] Nagai, K., Jiang, M. H., Hada, J., Nagata, T., Yajima, Y., Yamamoto, S., Nishizaki, T. 2002. (-)-Epigallocatechin gallate protects against NO stress-induced neuronal damage after ischemia by acting as an antioxidant. *Brain Res.* 956, 319-322.

[41] Nair, M. P., Mahajan, S., Reynolds, J. L., Aalinkeel, R., Nair, H., Schwartz, S. A., Kandaswani, C. 2006. The Flavonoid Quercetin Inhibits Proinflammatory Cytokine (Tumor Necrosis Factor Alpha) Gene Expression in Normal Peripheral Blood Mononuclear Cells via Modulation of the NF-κβ System. *Clin. Vaccine Inmunol.* 13(3), 319-328.

[42] Nicholson, S. K., Tucker, G. A., Brameld, J. M. 2008. Effects of dietary polyphenols on gene expression in human vascular endothelial cells. *Proc. Nutr. Soc.* 67, 42-47.

[43] Nonaka, T., Akimoto, T., Mitsuhashi, N., Tamaki, Y., Yokota, S., Nakano, T. 2003. Changes in the localization of heat shock protein 72 correlated with development of thermotolerance in human esophageal cancer cell line. *Anticancer Res.* 23, 4677-4687.

[44] Nishinaka, T., Ichijo, Y., Ito, M., Kimura, M., Katsuyama, M., Iwata, K., Miura, T., Terada, T., Yabe-Nishimura, C. 2007. Curcumin activates human glutathione *S*-transferase P1 expression through antioxidant response element. *Toxicol. Lett.* 170, 238-247.

[45] Menon, V.P., Sudheer, A.R. 2007. Antioxidant and anti-inflammatory properties of curcumin. *Adv. Exp. Med. Biol.* 595, 105-125.

[46] Motterlini, R., Foresti, R., Bassi, R., Green, C.J. 2000. Curcumin, an antioxidant and anti-inflammatory agent, induces heme oxygenase- 1 and protects endothelial cells against oxidative stress. *Free Radical Biol. Med.* 28, 1303-1312.

[47] Okabe, S., Fujimoto, N., Sueoka, N., Suganuma, M., Fujiki H. 2001. Modulation of gene expression by (2)-Epigallocatechin Gallate in PC-9 Cells using a cdna expression array. *Biol. Pharm. Bull.* 24(8), 883-886.

[48] Ohnishi, K., Takahashi, A., Yokota, S., Ohnishi, T. 2004. Effects of a heat shock protein inhibitor KNK437 on heat sensitivity and heat tolerance in human squamous cell carcinoma cell lines differing in p53 status. *Int. J. Radiat. Biol.* 80, 607-614.

[49] Pitozzi, V., Jacomelli, M., Zaid, M., Luceri, C., Bigagli, E., Lodovici, M., Ghelardini, C., Vivoli, E., Norcini, M., Gianfriddo, M., Esposto, S., Servili, M., Morozzi, G., Baldi, E., Bucherelli, C., Dolara, P., Giovannelli, L. 2010. Effects of dietary extra-virgin olive oil on behaviour and brain biochemical parameters in ageing rats. *Brit. J. Nutr.* 103, 1674-1683.

[50] Plaumann, B., Fritsche, M., Rimpler, H., Brandner, G., Hess, R. D. 1996. Flavonoids activate wild-type p53. *Oncogene* 13:1605-1614.

[51] Ramirez, M. R., Izquierdo, I., Raseira, M. C. B., Zuanazzi, J. A. S., Barros, D., Henriques, A.T. 2005. Effect of lyophilised *Vaccinium* berries on memory, anxiety and locomotion in adult rats. *Pharmacol. Res.* 52, 457-462.

[52] Ramirez, M. R., Guterres, L., Dickel, O. E., Castro, M. R., Henriques, A. T., de Souza, M. M., Marti-Barros, D. 2010. Preliminary studies on the antinociceptive activity of *Vaccinium ashei* berry in experimental animal models. *J. Med Food.* 13, 1-7.

[53] Ramirez, M. R., Apel, M. A., Raseira, M. C. B., Zuanazzi, J. A. S., Henriques, A. T. 2011. Polyphenols content and evaluation of antichemotactic, antiedematogenic and antioxidant activities of *Rubus* sp. Cultivars. *J. Food Biochem.* 35,1389-1397.

[54] Ramirez, M. R., Schnorr, C.E., Feistauer, L.B., Apel, M., Henriques, A.T., Moreira, J.C.F., Zuanazzi, J.A.S. 2012. Evaluation of the polyphenolic content, anti-Inflammatory and antioxidant activities of total extract from *Eugenia pyriformes* Cambess (Uvaia) Fruits. *J. Food Biochem.* 36, 405-412.

[55] Ramirez, M. R., Zanchin, N.I.T., Henriques, A.T., Zuanazzi, J.A.S. 2012. Study of the effects of *Psidium cattleyanum* on gene expression from senescent mouse hippocampus. *BLACPMA.* 11 (2), 127-137.

[56] Rao, C.V. Regulation of COX and LOX by curcumin. 2007. *Adv. Exp. Med. Biol.* 595, 213-226.

[57] Rusak, G., Gutzeit, J., Ludwig, M. 2002. Effects of structurally relted flavonoids on hsp gene expression in human promyeloid leukaemia cells. *Food Technol. Biotechnol.* 40, 267-273.

[58] Sandur, S.K., Ichikawa, H., Pandey, M.K., Kunnumakkara, A.B., Sung, B., Sethi, G., Aggarwal, B.B. 2007. Role of pro-oxidants and antioxidants in the anti-inflammatory and apoptotic effects of curcumin (diferuloylmethane). *Free Radical Biol. Med.* 43, 568-580.

[59] Savaskan, E., Olivieri, G., Meier, F., Seifritz, E., Wirz-Justice, A., Muller-Spahn, F. 2003. Red wine ingredient resveratrol protects from βamyloid neurotoxicity. *Gerontology.* 49, 380-383.

[60] Shafiee-Kermani, F., Grusak, M. A., Gustafson, S. J, Lila, M. A., Niculescu, M.D. 2013. Lower concentrations of blueberry polyphenolic-rich extract differentially alter HepG2 cell proliferation and expression of genes related to cell-cycle, oxidation and epigenetic machinery. *J. Nutr. Disorders Ther.* 3, 1-7.

[61] Shishodia, S., Singh, T., Chaturvedi, M. M. 2007. Modulation of transcription factors by curcumin. *Adv. Exp. Med. Biol.* 595, 127-148.

[62] Singh, S., Aggarwal, B.B. 1995. Activation of transcription factor NF-κ B is suppressed by curcumin (diferuloylmethane). *J. Biol. Chem.* 270, 24995-25000.

[63] Soleas, G. J. Goldberg, D. M., Grass, L., Levesque, M., Diamandis, E. P. 2001. Do wine polyphenols modulate p53 gene expression in human cancer cell lines? *Clin. Biochem.* 34, 415-420.

[64] Sonja, K., Nicholson, G., Tucker, A., Brameld, J.M. 2010. Physiological concentrations of dietary polyphenols regulate vascular endothelial cell expression of genes important in cardiovascular health. *Brit. J. Nutr.* 103, 1398-1403.

[65] Sun, Q., Cong, R., Yan, H., Gu, H., Zeng, Y., Liu, N., Chen, J. and Wang, B. 2009. Genistein inhibits growth of human uveal melanoma cells and affects microRNA-27a and target gene expression. *Oncol. Rep.* 22, 563-567.

[66] Sutherland, B. A., Rahman, R.M., Appleton, I. 2006. Mechanisms of action of green tea catechins, with a focus on ischemia-induced neurodegeneration. *J. Nutr. Biochem.* 17, 291-306.

[67] Tedeschi, E., Menegazzi, M., Yao, Y., Suzuki, H., Forstermann, U., Kleinert, H. 2004. Green tea inhibits human inducible nitric-oxide synthase expression by down-regulating signal transducer and activator of transcription-1R activation. *Mol. Pharmacol.* 65, 111-120.

[68] Veeriah, S., Kautenburger, T., Habermann, N., Sauer, J., Dietrich, H., Will, F., Pool-Zobel, B.L. 2006. Apple Flavonoids Inhibit Growth of HT29 Human Colon Cancer Cells and Modulate Expression of Genes Involved in the Biotransformation of Xenobiotics. *Mol. Carcinog.* 45, 164-174.

[69] Wei, Q. Y., Chen, W. F., Zhou, B., Yang, L., Liu, Z. L. 2006. Inhibition of lipid peroxidation and protein oxidation in rat liver mitochondria by curcumin and its analogues. *Biochim. Biophys. Acta.* 1760, 70-77.

[70] Weinreb, O., Mandel, S., Amit, T., Youdim, M.B. 2004. Neurological mechanisms of green tea polyphenols in Alzheimer's and Parkinson's diseases. *J. Nutr. Biochem.* 15, 506-516.

[71] Yokota, S., Kitahara, M., Nagata, K. 2000. Benzylidene lactam compound, KNK437, a novel inhibitor of acquisition of thermotolerance and heat shock protein induction in human colon carcinoma cells. *Cancer Res.* 60, 2942-2948.

[72] Yuan, J., Koh, W., Sun, C., Lee, H., Yu, M. C. 2005. Green tea intake, ACE gene polymorphism and breast cancer risk among Chinese women in Singapore. *Carcinogenesis.* 26, 8, 1389-1394.

[73] Zhuang, H., Kim, Y. S., Koehler, R. C., Dore, S. 2003. Potential mechanism by which resveratrol, a red wine constituent, protects neurons. *Ann. N. Y. Acad. Sci.* 993, 276-286.

[74] Zhang, X. G., Xu, P., Liu, Q., Yu, C. H., Zhang, Y., Chen, S. H., Li, Y. M. 2006. Effect of tea polyphenol on cytokine gene expression in rats with alcoholic liver disease. *Hepatobiliary Pancreat. Dis. Int.* 5, 268-272.

BIOGRAPHICAL SKETCH

Maria Rosana Ramirez
National Council of Scientific and Technical Research, Argentina.

Education: PhD in Pharmaceutical Sciences.

Business Address: Monseñor Tabella 1450, Concordia, ER, Argentina

Research and Professional Experience: Functional foods, Pharmafood, Experimental Biology.

Professional Appointments: Adjoint investigator.

Honors: IBRO, Ricardo Miledi Neuroscience Training Program (Grass Foundation, SFN).

Publications: www.conicet.gov.ar.

INDEX

B

C

D

E

F

G

H

Q

R

reactions, 23, 59, 64, 68, 119
reactive oxygen, 23, 31, 64, 68, 91, 100,
    116, 129
reactivity, 117
reception, 5
receptor, 100, 102, 103, 133, 134, 137, 146,
    148
recovery, 10, 33
rectocele, 92
red blood cells, 117
red wine, 129, 142, 146, 152
regeneration, 23, 41, 53, 144
regions of the world, viii, 57, 90
regression, 138
relevance, viii, ix, 22, 35, 90
relief, 93
remodelling, 98
renin, 67
repair, 33, 41
repression, 103, 132
researchers, 141, 142
residue, 24
resistance, 29, 32, 37, 103, 132, 133
resolution, 93, 119
response, 2, 17, 38, 43, 66, 71, 83, 93, 97,
    98, 99, 102, 104, 113, 116, 128, 129,
    131, 133, 135, 136, 137, 144, 149, 150
restoration, 149
restructuring, 42
resveratrol, viii, 22, 29, 129, 136, 143, 146,
    147, 148, 151, 152
rheumatoid arthritis, 31, 93, 106, 111, 117,
    121
rhizome, 27, 36, 129
rings, 92
risk, vii, 2, 26, 39, 59, 65, 71, 74, 103, 105,
    134, 135, 136, 148, 152
rodents, 33, 36, 46, 139
root, 60, 66, 68, 76, 85, 129

S

safety, 44, 49, 113, 115
saturation, 132
science, 50

scientific papers, 54
scientific publications, viii, 21
scope, 59
SDS, 53
seafood, 77
secondary metabolism, 61
secretion, 66, 68, 114, 129, 136
seed, 27, 29, 46, 50, 52, 60, 69, 75, 76, 78,
    84, 103, 115, 122, 143
selectivity, 7, 16
selenium, 61, 134, 148
self-assembly, 39
sensing, 136
sensitivity, 2, 145, 150
Serbia, 21, 53, 54
Sericoside Phytosome®, 42, 45
serotonin, 71
serum, 26, 28, 29, 34, 35, 44, 69, 73, 98, 99,
    100, 101, 102, 103, 104, 105, 107, 112,
    113, 121
serum albumin, 104
serum ferritin, 44
sex, 95, 105
sex steroid, 95
shade, 60
shock, 131, 132, 147, 148, 149
short-term memory, 28
showing, 91, 111, 131, 141
side chain, 73, 91
side effects, 38, 67, 72, 74
signal transduction, x, 65, 69, 128, 141, 142
signaling pathway, 83, 129, 135
signalling, 82, 93, 124
signs, 116
Siliphos®, 27, 33, 42
Silipide®, 33
Silymarin Phytosome®, 27, 30, 31, 32, 36,
    42, 45, 47, 48
simulation, vii, 2, 9, 52
Singapore, 152
skeleton, 65
skin, 23, 28, 30, 31, 37, 41, 42, 43, 44, 48,
    49, 50, 60, 81, 112
skin cancer, 48
skin diseases, 42, 44

T

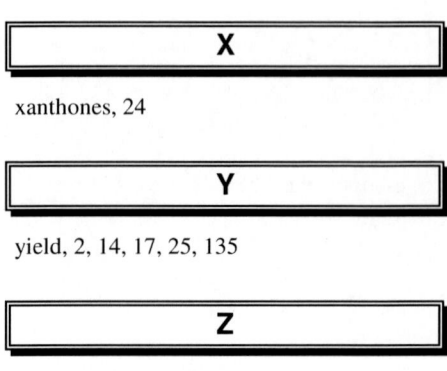